THE NEW FORESTER

Berry van Gelder

Phil O'Keefe

Practical
ACTION
PUBLISHING

Intermediate Technology Publications Ltd
trading as
Practical Action Publishing Ltd
27a Albert Street, Rugby, CV21 2SG, Warwickshire, UK
www.practicalactionpublishing.org

© Intermediate Technology Publications Ltd, 1995

First published in 1995
Transferred to digital printing in 2008
ISBN 978 1 85339 2320

A catalogue record for this book is available from the British Library.

Since 1974, Practical Action Publishing has published and disseminated
books and information in support of international development work
throughout the world. Practical Action Publishing is a trading name
of Practical Action Publishing Ltd (Company Reg. No. 1159018), the
wholly owned publishing company of Practical Action. Practical Action
Publishing trades only in support of its parent charity objectives and any
profits are covenanted back to Practical Action (Charity Reg. No. 247257,
Group VAT Registration No. 880 9924 76).

Typeset by Gary Hailey

Contents

Boxes

Foreword

There is a growing literature devoted to the role of trees in developing countries. At the present moment, much of this literature simply describes present patterns of wood use. What is needed is a guide to help design rural forestry projects that provide the wood products that local people need.

This volume describes the authors' experience in a variety of wood projects. The central message of the volume is that there is urgent need to develop a new kind of forestry — and therefore train a new type of forester — if the issue of wood provision for rural people is to be addressed. Our experience drives us towards a search for a new methodology in forestry that builds upon local knowledge and local structures.

This search for a new methodology is simultaneously humiliating and liberating. It is humiliating in that we have to confess that our traditional training as professionals limits our ability to design successful wood projects: we look back on landscapes littered with project failures from ourselves and others. It is liberating because we are beginning to learn from our failure. The single biggest lesson is, again, simultaneously humiliating and liberating. We do not have, nor can anyone else have, a simple, single technical answer. As a consequence, we have explored a new way of seeing solutions building from wood use to local landscape. It is this new way of seeing that is shared in this volume.

The volume is designed, with boxes, so that those leading courses in rural forestry can beg, steal and borrow material from the book. It is meant to be used, rejected or rewritten by anyone committed to building sustainable livelihood and thus landscape systems. We look forward to the critique from other fieldworkers and, hopefully, to an account of their success in building forestry projects.

There are many people to thank. First, the farmers themselves who forgave our ignorance and helped us to understand. Second, the Dutch Foreign Ministry who, under its aid programme, continue to fund innovative work. Third, colleagues at the ETC Foundation who share a vision, if not always perfect practice, of participatory development. And finally the next generation who provided much material input for this volume: Arend Jan van Bodegem, Nicki Allsop and Chris Howorth.

Phil O'Keefe   *Berry van Gelder*
ETC (UK)   *FMD Consultants*
117 Norfolk Street   *P.O. Box 10363*
North Shield   *7301 GJ Apeldoorn*
Tyne and Wear   *The Netherlands*

Preface

This book would not have been possible without the continuous support of the environment programme of the Dutch Ministry of Foreign Affairs. Particular thanks are due to Peter Lammers and Paul Hassing who have actively led policy debates around the issues of energy and environment at a global level. It is on the back of such debates that we have begun to understand what local people themselves can do in forestry. Many times, understanding what local people can and are doing is an extremely sobering experience — it questions our conventional view of the delivery of development and our unfounded assumption that local people know nothing. If this book helps to let other people question their own assumptions and understand what local farmers do as everyday practice, then it goes some way to accelerating a new kind of forestry

Phil O'Keefe
ETC (UK)
January 1994

1. Setting the Scene

The challenge of wood: the biomass economy

Wood is everywhere; it is the real wealth of nations. Wood is more than trees and includes shrubs and other cellulose producing plants. Woody biomass in general, not trees in particular, is the total wealth.

Woody biomass still dominates most of the earth's surface, although it has been quite rapidly removed. Two recent studies, reported by the United States Agency for International Development, conducted by the Food and Agriculture Organization (FAO) and the World Resources Institute (WRI), suggest that deforestation rates are approximately 50 per cent higher than was previously presumed. Crudely, FAO estimates that 42 million acres a year are being deforested and the WRI estimate gives a somewhat higher rate of 50 million acres a year. The real wealth of nations is under threat.

This threat to the real wealth of nations is not just a threat to the environment. It is true that continuing deforestation causes soil erosion and land degradation. It is also true that the loss of forests is the loss of a common property resource, not just for immediate consumption, but a common property resource that acts, among other things, as a carbon sink to absorb the major greenhouse gas, carbon dioxide. The problem is more than an environmental problem — it is a problem of basic needs provision, a problem of subsistence survival.

In general, subsistence economies in developing countries are biomass economies. Of course, the plastic, glass, steel and concrete goods of mass production are making inroads but household economies essentially remain biomass based. Housing itself, energy, furniture and utensils are still biomass products. The yield of trees and shrubs provides thatch, fodder and a host of other products which serve each family household. Woody biomass yields food for people and animals. Deforestation is thus an erosion of local entitlement to subsistence resources as well as being an environmental problem that eradicates local, national and global common property resources.

There is clearly a demand for the products of tropical woody biomass. High-value clothing, for example good quality cotton and silk, is premised on the existence of woody biomass. Beyond that, rare tropical woods command high prices in custom design furniture for the developed world. These commercial demands, which are increasingly profitable and therefore attractive investments, are insufficient to address the problem of deforestation. Quite simply, reversing the decline in basic needs provision in the biomass economy and strengthening ways of sustaining common property resources is a challenge that can be met only by starting with the majority of consumers.

The majority of consumers are not the rich, with large, disposable incomes, but the world's poor. It is this poor population, especially the individual farm household, that has to be the starting point for a global effort. The first hurdle in this effort is to convince professionals that the farmers' way of seeing both the problems and opportunities contains the ingredients for successful biomass programmes. Encouraging farmers to be the generators of the real wealth of the nation is the easiest technique of developing woody biomass, because they live in, with and by woody biomass.

In many cases, this need not be difficult. Local people already actively manage and experiment with local woody biomass. But it can be a difficult problem because professionals are not trained to see that local people actively create and manage their own environments. It is not only foresters that are not able to see the reality in front of their eyes; rural development professionals, in general, see only the technical solutions. At a local level, they will count a farmer as rich if the farmer has ten cows; they do not count the farmer rich if the farmer has one hundred trees.

Foresters, and other rural development experts, are technically myopic. It is not only the use of Latin nomenclature, of physical production targets and the encouragement of monocultural plantation; it is not even that they act as champions and statesmen for specific species rather than creating sustainable land-use systems; it is that they will not work with what people are already doing. Quite simply, foresters must start with people. They should not begin by asking for how much land is available for planting but how many people are available to plant.

Starting with people will quickly show that constraints are frequently not local but external. Changes in land tenure, encroachment on local common property resources, lower wood prices and the collapse of traditional institutions of resource control and arbitration establish a framework which is superimposed on local opportunity. The challenge for foresters, and other rural development professionals, is to mitigate the negative external influences, which might require substantial policy changes at national level, while simultaneously building on the positive local responses that already exist.

Hopefully this book will contribute by helping professionals see new opportunities by working with, not against, local people.

From plantation to fields: location of the problem

To find a solution, the problem must be accurately located. The failure of any intervention is usually to be found in the initial misperception of the problem. If the basis of a solution (i.e. the location of the problem) is wrong, it obviously follows that any intervention will be misdirected.

Most forestry interventions over past fifteen to twenty years have viewed the biomass problem as a simple supply and demand problem. It was argued that people were extracting more biomass than the environment could produce in a sustainable manner. Viewed from this perspective, the solution was self evident: if projected biomass demands exceeded supplies, one either planted more trees to shift the supply curve upwards or devised policies to reduce demand and shift the demand curve downward. As a result, foresters have tried to increase tree supplies with various large scale approaches such as monoculture plantations, peri-urban woodlots, community woodlots and the increased policing of forests and woodlands. Their rationale has been to plant as many trees as quickly as possible. Unfortunately, and all too often, decisions to spend large sums of money planting trees have been taken without considering other options and the consequences of existing market and policy failures. Foresters have only themselves to blame for excluding options, for courting failure.

Fortunately, a re-thinking of the biomass problem is underway. This is the result of a number of factors, including the obvious failure of previous approaches, the demand for more appropriate interventions at local level and the further reluctance of donors to finance future projects until a stronger range of interventions is available.

During the 1970s and 1980s, opinion about the role and responsibility of forestry in the context of rural development changed dramatically. Forestry moved from the traditional orientation of forest protection and the exclusive encouragement of industrial wood production in the interests of national economic growth, to embrace far broader economic, social and environmental concerns. For example, there can be few foresters who are unaware of the implications posed by current patterns of deforestation to such global-scale environmental issues as global warming and, with particular reference to tropical moist forests, the question of the loss of biodiversity and genetic resources. In addition, at regional and local scales, there is growing recognition of the vital role that trees, and indeed all forms of woody biomass, play in sustaining traditional rural economies. Foresters now know the links between the degradation of woody biomass resources and the issue of accelerating rural poverty.

These changes reflect broad shifts in global perceptions and expectations which have been given a common focus in the recent Brundtland Commission Report (World Commission on Environment and Development). The call is now for a form of forestry which will contribute to the process of sustainable development; a development that is equitable and which meets the needs of the present without compromising the ability of future generations to meet their own needs. Implicit in this definition is a need for new forestry initiatives which contribute to a participatory, equitable, decentralized and self-sustaining process of rural development throughout the developing world.

In the rest of this chapter, the changes in forestry orientation are explored. The broad aim is to show the vital role of woody biomass as an integral component of rural development, extending beyond the provision of industrial raw materials. To this end, we journey out of the forest, be it natural or plantation, to the large, but widely dispersed rural tree resource (natural woodlands and trees and shrubs 'on farm') which, until relatively recently, has been largely ignored by foresters, agriculturalists and others involved in rural development. The groundwork is laid for a 'new way of seeing' trees, and indeed all woody biomass, within rural landscapes. The multiple role this important renewable resource plays in rural economies, and hence development, is emphasized; woody biomass problems are given a new perspective and, more importantly, we highlight a new orientation for forestry which is vital if sustainable production initiatives are to be implemented.

It is important at this stage, even at the risk of repetition, to correct three popular misconceptions about deforestation and woodfuel:

- The common assumption that deforestation is caused by commercial logging or cutting for woodfuel is not true; agricultural colonization is the major cause of deforestation.
- The presumption that forests are the primary source of woodfuel for rural people is quite simply wrong. Generally speaking, where there are people there are no forests, and vice versa. In other words, whilst over 90 per cent of

woody biomass in the Third World is in forest areas, 90 per cent of biomass fuels come from the agricultural landscape. It is trees outside the forest which are the vital source of fuelwood in developing countries.
- The notion that rural people fell trees for domestic energy use is again wrong; woodfuel is a residue of the other end uses of wood in the local economy. Quite simply, woodfuel is what is left over.

Two points arise from this which have important implications for our perception of biomass problems and hence social forestry initiatives. First, as biomass is produced and consumed at the local level, its use can only be understood at the local level. Second, for rural people there can be no such thing as a single purpose tree, although this is not to say that people do not prefer certain species for a particular end-use.

The traditional orientation of forestry in the context of development has focused on its macro-economic function and largely ignored the vital contribution of all woody biomass resources (natural woodlands, dispersed trees, scrub, hedges, crop residues and so on) to local rural economies. Despite this neglect by professionals, rural people themselves have always recognized woody biomass as a multi-purpose resource providing a large range of products and services.

The range of uses and functions of woody biomass in any rural setting is enormous. There are many productive functions, from the provision of building materials to medicines, and service functions such as environmental protection and ornamental provision. Woody biomass is part and parcel of every livelihood system. In fact, traditional economies can easily be defined as biomass economies. Box 1 gives an overview of the most important of its functions.

Although the identification of the multiple functions of woody biomass is possible, the complex role that trees can play within agricultural production systems is underestimated. Information about tree species on farms is still scarce. The amount of woody biomass vegetation in rural areas is often underestimated. It is scattered and, because it varies in structure, is not always conspicuous.

For example, two major biomass evaluations have been made in Mozambique. The first was an inventory of forestry resources in Mozambique carried out, in 1979-80, with FAO assistance. It was based on satellite imagery and aerial photographs taken in 1972. According to this survey, forests in Mozambique covered about 56.5 million hectares in 1972, equivalent to about 71 per cent of the country's area, with a total growing stock of about 2 billion cubic metres. A second biomass assessment, prepared in 1987 by ETC Foundation based on satellite remote sensing, indicated that Mozambique has a total growing stock of some 4 billion cubic metres. The major reason for the difference in assessment of total growing stock between the two surveys is that the first survey concentrated on the forests, while the second survey covered woody biomass over the whole of Mozambique. Total annual demand for fuelwood is, however, only 30 million cubic metres, clearly an insubstantial inroad into the total resource. But woody biomass shortages are emerging, particularly around the urban and peri-urban areas, concentrated in Maputo and Nampula provinces where most people live. Shortages, and thus the need to domesticate woody biomass production, do exist despite the overall woody biomass resource abundance.

Box 2 provides details of woody vegetation on farms in Kenya. Although the scale of analysis has altered to a local level, more than fifteen per cent of all land is still under forest cover, much of it planted. These particular areas in Kenya are high-

Box 1 Functions of woody biomass in rural areas

Food
For home consumption or for sale:
* Fruits
* Nuts
* Spices, and
* Leaves.

Fodder
For year-round or seasonal feeding
of livestock, either in grazing or
cut and carry systems:
* Leaves
* Pods, and
* Twigs.

Construction material
For houses, fences and granaries:
* Sawn timber
* Split wood
* Posts and poles
* Branches
* Twigs, and
* Leaves.

**Domestic utensils and
agricultural implements**
For furniture, utensils, tool
handles and ploughs:
* Chairs
* Tables
* Beds, and
* Bowls.

Fuel
For domestic purposes (cooking,
heating), for specialist uses or for sale:
* Firewood (twigs, roots, branches,
  split wood, leaves), and
* Charcoal.

Income
* Sale of tree products, and
* Tree as 'savings account'.

Medicines
* Leaves
* Bark
* Sap
* Roots
* Fruits, and
* Pods.

Social purposes
* Religious or ceremonial
  purposes
* Ornamental purposes, and
* Social fire-places.

Demarcation and fencing
* Trees or shrubs for land
  demarcation
* Living fences
* Fencing materials.

Environmental functions
* Soil fertility improvement
* Soil erosion control
* Water conservation
* Wind shelter
* Shading, and
* Weed control.

Raw materials
For home industries or for
sale:
* Dyes
* Fibres
* Honey, beeswax
* Oils and resins
* Wood for craft purposes
* Fruits, nuts and the like for
  processing, and
* Tannins.

potential farming areas and descriptions of existing land use significantly omit mention of woody biomass. Many rural development professionals pay little attention to the potential supporting role of trees and shrubs towards crops, livestock or sustainable environmental practices. Even foresters tend to overlook anything that does not look like a real forest, such as single trees, shrubs, hedges or bushes.

Historically, little has been known about the problems of rural people maintaining access to tree products such as firewood, building material, fruits and fodder, or the environmental consequences of forest clearing, over cutting and over grazing. It has become clear that these problems are fairly new in most regions. They are not the result of irresponsible exploitation of forests and woodland, but of rapidly increasing demand for agricultural land in response to external influences. If agricultural colonization is the most important cause of the deforestation problem, then those colonizing land must be encouraged to plant woody biomass.

Tree planting is often seen as the solution to this problem. Foresters are expected to take the lead in rural tree planting efforts. However, within rural development circles, there are conflicting views on how to proceed. In particular, the role of rural people in tree planting is unclear.

As woody biomass resources are gathered freely from the local environment, any attempt to understand their use requires knowledge of the local biomass resource base. The woody biomass resource itself reflects an area's land use pattern. This, in turn, is a reflection of both the local environment and the land management system.

As forest resources have declined over time, primarily as a result of land clearance for agricultural production, rural communities in the developing world have recognized the importance of having woody biomass vegetation near their villages to provide the goods and services central to their livelihood systems. These resources have therefore been progressively incorporated and maintained within local farming systems through indigenous agroforestry. This occurred across a wide spectrum of agro-ecological land use and population density situations — resulting in a great diversity of indigenous agroforestry systems.

Most rural landscapes contain substantial quantities of on farm woody biomass vegetation which constitutes an integral part of the farming system. Similarly, trees and shrubs occur widely over areas of common property: along river banks or road sides, over grazing lands, steep hill slopes, within open areas of villages and so on. Common property resources are frequently a major source of wood and tree products. They constitute a major component of the overall agricultural system, filling gaps in the resource and income flows and provide complementary inputs to maintain the agricultural and household systems. Common property resources are often very important to the most disadvantaged sectors of a local rural community such as women, the poor, the landless and minority ethnic or religious groups.

To understand woody biomass resource use we must understand the local resource base. On its own, however, this is frequently not enough. The key issue is not the existence of biomass resources, but their availability to local people for different end uses. Access, which reflects traditional patterns of use, is the key. A series of factors limit access to these resources. Constraints affect different sections of the community in different ways. These constraints reflect features of the local physical environment and the existing social and economic relations within a community. Therefore, the form constraints take is highly locality-specific. These constraints must be seen against people's expressed entitlement to local resources. Understanding constraints and expressed entitlement is vital in order to understand

Box 2 Woody vegetation on farms in Kenya

The Kenyan Highlands provide a good example of woody biomass in the landscape. Land use, in these densely populated areas, is extremely intensive but trees are not recorded in land use statistics. Yet the landscape is full of woody elements, such as woodlots, windbreaks and scattered trees in cropland.
   The main types of woody vegetation are:

- **Woodlots**: Small mini-plantations with fast-growing trees, such as eucalyptus or black wattle (*Acacia mearnsii*)
- **Windrows**: Narrow rows of trees along the fields or on the roadside; common species are cypress (*Cupressus lusitanica*), *Greville's robusta* and eucalyptus
- **Trees in agricultural land**: Individual or small groups of trees growing within fields or in grazing land, usually deliberately planted, sometimes grown naturally; for example, *Sesbania sesban, Markhamia luteax, Croton spp.*
- **Trees around homesteads**: Fruit or shade trees, sometimes timber trees, growing on the compound near the houses; for example, mango (*Mangifera indica*), avocado (*Persea americana*), loquat (*Eriobotrya japonica*)
- **Bushes**: Natural shrubland, sometimes with some taller trees; it varies in density and height according to age and intensity of cutting or grazing, and
- **Forest**: Remnants of the original natural vegetation, dominated by tall trees, but varying in density and often found on steep slopes or along rivers.

The general pattern is that, in the most densely populated areas, the proportion of man-made vegetation types (woodlots, windrows, hedges, trees in cropland, trees on compounds) is high and the area under natural vegetation is low. In less densely populated regions, the opposite will be found since both the pressure on the land and wood resources are lower. The table below outlines percentage wood cover in these districts and breaks down the different types wood occurrence.

Woody vegetation in three districts

District areas (km²)	3500	2200	2500
Woody vegetation (of which land covered)	16.5%	17.5%	15%

Breakdown of woody vegetation (%)

Planted (total):	41	62	39
Woodlots	13	20	18
Windrows	6	3	2
Hedges	12	32	5
Trees in agricultural land	6	6	12
Trees on compounds	4	1	4
Natural (total):	59	38	59
Bushes	32	23	29
Forest and woodland	27	15	30

the ways in which woody biomass can be produced and used within a local rural economy.

Key constraints are those that limit access to resources. Three broad categories of access constraints can be identified in rural areas. These are:

- The limitations imposed by the location of the resources in relation to demand
- The issue of land and tree tenure, and
- The ways the local biomass resource base is managed.

Where resources remain locally abundant and population densities are low, it is often unnecessary for local people to actively manage trees, although some form of extensive management system exists. However, in an ever increasing number of situations, rural people need assistance to maintain the land use systems they need.

It is important to recognize that successful traditional tree farming systems in the developing world were responses to changes which were occurring slowly. Today, faced with rapid change, rural people are faced with situations they have not previously encountered. In many cases they have reached the point where they cannot sustain their livelihood systems within the framework of their existing knowledge, resources or institutions. In a wide range of situations, rural people need assistance to maintain the sustainability of their local biomass economy.

Foresters have an important role to play in helping to sustain the biomass economies of rural areas. However, they can only perform this role successfully if they are willing to come out of the forest, listen to, and work with, rural people and other institutional bodies involved in rural development.

From fuelwood to trees: redefining intervention

Traditional forestry

Until the early 1970s, most forestry development reflected the traditional macro-economic focus. It followed a northern model of large-scale industrial-commercial plantations and natural forest management to earn revenue and foreign exchange. For many countries in Africa and Asia, this dates back to the colonial era, a century or more ago, when imperial powers established forestry departments as integral parts of their colonial administrative structures to serve imperial priorities. For example, the forest department was one of the first departments created in Sudan by the British Government, in 1901, primarily to secure a good supply of wood for the steamers plying the River Nile. In Ghana, then the Gold Coast, the forestry department was established in 1908. Its remit was to facilitate and control the extraction of forest products, primarily wood. For the colonial forester, forests were, for the most part, regarded as the source of one product: wood. Extracting timber as fast as possible was the single goal. The interdependence between forests and local people was largely ignored, as were trees outside the forest.

Antagonism between rural people and foresters grew as land was expropriated by the State to create State forests. Foresters assumed the role of policemen, guarding the national resource from so-called illegal exploitation. Foresters were also tax collectors, raising revenue from fines and cutting permits. For example, in Mali, the forestry code established by the French colonial administration was essentially restrictive and punitive. Wood fallow land (considered vacant and without ownership) was expropriated to create State forests with little regard for the existence of customary land-rights. The system of fines imposed was seen to be both abusive and random. Conflict between foresters and local people rapidly developed, particularly between foresters and pastoralists. Pastoralists move herds through land resources, which never sat well with a land take associated with traditional forestry.

On independence, forestry departments in both Africa and Asia inherited much of the infrastructure and, more importantly, the ideologies and attitudes of the colonial offices they replaced. The belief that forests were of value primarily for wood production, in a macro-economic context, continued to prevail. This attitude was compounded by the forest policies of international donor agencies who were now involved in supplying forestry aid to developing countries. Agencies' policies reflected the industry-led theories and programmes of development which predominated during the 1950s and 1960s. For example, the FAO Division of Forestry and Forest Products (now the Forestry Department) described the forest as:

essentially a wood-producing unit. ...its treatment must be conditioned by the technological properties of its products for their industrial utilization.

This industrial emphasis did not deny the protective environmental role of forests. In many countries, forest reserves were created to preserve catchment areas and water supplies. However, priorities in budget allocation and expertise continued to be directed at strengthening the capacity of forestry departments to manage forests and prevent encroachment. The aim was to expand the production of wood for export and as an input to the emerging wood-based industries, to supply saw and pulp mills. Forestry remained a largely technical discipline. Foresters continued in their traditional, but often contradictory roles, of protector and exploiter of trees within both natural and industrial-plantation forests.

The turning point came in the mid 1970s. By the end of the decade community forestry, or trees for the people, were the buzzwords on everyone's lips. This radical shift in forestry thinking stemmed from a number of underlying developments.

First, there was the discovery of the fuelwood crisis. The first oil price hike of 1973 had put energy issues to centre stage on the international political agenda. Suddenly attention was drawn to the dependence of some 2 500 million people, or roughly half of the world's population, on firewood, charcoal and other biomass fuels for cooking and other essential energy needs. For the first time, national energy balances of supply and demand began to be constructed to include fuelwood. Playing this numbers game, planners came up against the fuelwood gap between supply and demand that, in the absence of intervention, would widen considerably. Unfortunately, the foresters only counted trees inside the forest — the one woody biomass resource rarely used for woodfuel. Viewed from the perspective of these simple models, answers appeared self-evident. If projected demand outstripped potential supplies, then the answer was to develop fuelwood plantations.

At the same time, industry-led theories of development had given way to a people-centred rural development focus. This stressed the need to enable rural populations to meet their basic needs. This shift served to highlight problems where local people were finding it difficult to acquire biomass to meet their energy requirements.

Finally, the accelerated loss of tree stocks which occurred throughout the Sahelian region of Africa during and after the 1968–73 droughts, heightened global concern over the environmental role of tree cover in sustaining soil and water quality and associated agricultural productivity.

Concern over these overlapping issues meant that developing world governments, international donor agencies and foresters could no longer ignore the relationships between trees and people. A series of studies and meetings held during the 1970s culminated in The Eighth World Forestry Congress in 1978 which took as its theme Forests and People. In the same year the FAO published Forestry in Rural Development in which community forestry was defined as:

...any situation which intimately involves local people in a forestry activity.

The FAO paper discussed a range of situations, from woodlots in areas short of wood and other forest products for local needs, to growing trees at farm level to provide cash crops and the processing of forest products at the household, artisan or small industry level to generate income, to the activities of forest dwelling communities. It excluded large-scale industrial forestry and any other form of forestry which contributes to community development solely through employment and wages. However, it included activities of forest industry enterprises and public forest services which encouraged and assisted forestry activities at the community level.

The World Bank, in their Forestry Sector Policy Paper (1978), acknowledged that past approaches to forestry had been inadequate and too narrowly focused to be of any real help meeting the needs of the urban and rural poor in the developing world. In the late 1970s, after 25 years of almost exclusive focus on the industrial value of forests and extensive technical assistance and investment programmes for pulp mills, saw mills and industrial plantation forestry, it was belatedly realized that such programmes were doing little for the vast majority of the rural and urban poor of the developing world.

The result was the launching of numerous community or social forestry programmes, many aimed at solving the fuelwood crisis, which took place with a rapidity unusual in the inertia-ridden world of development planning. International aid for such programmes rose from a negligible amount in the early 1970s to more than $500 million invested between 1977 and early 1984.

Despite such a radical shift in thinking and investment, the failure rate of social forestry interventions was unacceptably high. A noted commentator, Agarwal, describes a number of such failures, including projects in Ethiopia where labourers planted trees upside down, in Niger where saplings on communal grazing land were uprooted to allow cattle to graze, in India where eucalyptus saplings were uprooted by villagers and in the Philippines where Tinggian tribal communities deliberately started fires in areas taken over by the forestry department for community forestry. In one project, in Senegal, saplings distributed by foresters to local residents for planting on forestry lands, were purposefully destroyed by the residents. In Nepal, trees planted along irrigation channels were destroyed within two days of planting. In many villages of Tanzania, people were openly hostile to efforts to establish communal woodlots, uprooting or cutting the plants deliberately. In the Qala Nahal area of Kassala province, a community forest tractor driver was forced off the land by local people at gunpoint.

However, active resistance was unusual. More commonly, people tended to ignore such projects which consequently failed because of the lack of participation by the local community. For example, the Bangladesh Community Forest Project was set up in the early 1980s to provide fuelwood in north-west Bangladesh, an area which has one of the world's most acute rural energy crises. Woodlots were to be established in village communal lands, along road and canal sides and on government-owned land. Millions of seedlings were planted, but no attention was paid to their care. Survival rates of seedlings were less than five per cent.

Rural people have never seen forestry or tree planting as a separate activity from the rest of their production activities; the management of trees and shrubs is part and parcel of their land use strategy. But as most rural development projects are organized according to sectors, such as agriculture, forestry or animal husbandry where each sector ignores the contribution of the other within the same land use system, peoples' land use strategies have rarely been reflected in rural development initiatives. For example, while fuelwood shortages would be seen only as a forestry problem, declining soil fertility would be the concern of agriculturalists. Rural people make no such divisions.

This technical, sectoral approach fails to recognize the complexity of rural development problems. What this indicates is that a new approach to rural development is required. The approach must reassess the limited value of sectoral approaches, and reconsider the value of top-down, single-ministry initiatives. It is only through the adoption of an approach that starts from the existing entire production system and builds on knowledge skills and local knowledge that the biomass economy can be successfully sustained.

Consider the classical foresters. They have traditionally worked in the forest, usually under government control, and are responsible for the management and protection of natural forest or forest plantations in the interests of the State. Making the transition to social forestry is difficult because foresters cannot shake off the roles of manager and policeman. Despite much rhetoric to the contrary, foresters have retained their conventional approaches and familiar scientific methods. Many social forestry projects are merely scaled-down versions of the conventional industrial plantation forestry. Rows of a single species of trees (frequently non-indigenous fuelwood trees) are planted in straight lines and surrounded by fences. In many cases, the planning model has remained top-down, with government officials doing the planning. Community participation, as such, is confined to government experts informing people of their plans. Even where local people are involved, it is typically the economically and politically dominant men who control proceedings. Those who frequently have the greatest need of enhanced woody biomass supplies — women and the landless — remain excluded. Calls for people participation remain rhetoric contained in development agency reports and conference papers. People participation has rarely been reflected in actual activities on the ground.

Driven by the findings revealed by energy gap analysis, many early projects were designed by classical foresters. Fuelwood production was the main, and often sole, objective. These projects sought to secure fuelwood supplies for local people through the establishment of woodlots containing what professionals perceived to be fuelwood trees. But, fuelwood is only one of the many end uses of trees. More importantly, fuel provision may be only one of many problems rural people face, and is frequently not a priority concern. For example, in an area of Malawi which was experiencing growing fuelwood scarcity, a study found that the priority concern, as articulated by the community, was a shortage of construction timber; suitable poles for house building were even more difficult to find than fuelwood. Similarly, a study in Nepal showed fodder to be the principal concern. But classical foresters are still trying to solve the energy problem using fuelwood trees. Put simply, foresters had a solution in search of a problem. Therefore, it is hardly surprising that local support for forestry interventions of this type have been lacking.

Experience has also revealed fundamental flaws in the assessment of the fuelwood problem. In many cases, the fuelwood problem has been overstated. More importantly, the trees outside the forest, around homesteads, along road sides, within villages and on agricultural land, which are the primary source of fuelwood for rural people, have not been included in any analysis. Inventories of national wood stocks and yields concentrate on commercial forests and, to a lesser extent, natural woodlands. They neglected trees and other sources of woody biomass on-farm. In short, there is little knowledge of the contribution made from these sources to national wood stocks and yield. Not surprisingly, information on national wood stocks and yields is usually little more than guesswork. The tendency to approach local forestry interventions on the basis of national statistics is fundamentally flawed.

The premise that it is demand for fuelwood which creates deforestation is untrue. The primary cause of deforestation is land clearance for agriculture. One estimate suggests that between 1950 and 1983, this was responsible for seventy per cent of the permanent destruction of forests in Africa. These figures parallel global figures from the FAO which suggest that, especially in Africa, land clearance will be the major cause of forest clearance even in the next century. Failure to take account of the dynamics of land use is another fundamental flaw in the analysis of the wood problem.

In terms of remedial measures, the same mistakes have been made. The assumption that there has been one causative agent has led to a single remedial measure being employed. But there has been a multitude of reasons why there is a biomass problem. It follows that there should be a multi-sectoral, all-encompassing approach to that problem. Agroforestry is a potential approach.

Social forestry

As yet there has been no formal definition of agroforestry. Perhaps this defeats the objective of funding a new way of thinking about biomass systems. When specific guidelines are laid down, programmes can become too rigid and allow little room for manoeuvre.

A general definition of agroforestry, as currently practised, would include a recognition that it:

- Is a collective name for land use systems involving trees combined with crops or animals, or both, on the same unit of land
- Combines production of multiple outputs with protection of the resource base
- Places emphasis on the use of indigenous, multi-purpose trees and shrubs
- Is particularly suitable for low-input conditions and fragile environments
- Is more concerned with socio-cultural values than most other land use systems, and
- Is structurally and functionally more complex than monoculture.

This is a very scientific approach in which people are not directly identified. The emphasis on socio-cultural values and multi-purpose outputs raises the possibility of a people focus, but it is a muted voice. What is needed is a drive for agroforestry in which the social emphasizes that individuals and communities often know best and in which agroforestry emphasizes existing land use systems. Social forestry starts by assuming that the existing landscape contains clues of past, purposeful local action. The challenge is to discover with local people the clues that can help build a sustainable biomass economy.

2. Tools for Work — Indigenous Knowledge

Asking questions with people

Certain things must be clear when discussing the new approach to forestry in terms of rural development. It needs to be recognized that:

- People are the solution, not the problem, in forestry
- Local people define the opportunities with which they work
- Their stories contain the logic of their life, and
- Their actions are sensible.

For successful local development, conversation is better than administration. What is needed is not only a recognition of these things, but a commitment to applying them in forestry projects. The orientation should be towards the farmer and not towards the trees. The farmers' own experiences represent an enormous untapped resource which can be harvested to rural development efforts.

Many farmers and herders already practise a form of agroforestry, though it is not always recognized as such. The most widespread is the use of trees growing on their land. Most families have shade or fruit trees growing in their compounds. Many farmers foster useful seedlings that sprout simultaneously in their fields. Many grow tree and shrub crops, from oil palm and cocoa, to coffee and tea, banana or plantains.

Outside the compound or the tree-crop plot, most farmers have relied on natural processes of self-planting to provide them with trees. These processes are now under severe threat. As fallow periods are cut down, trees may no longer have time to reach maturity and start producing seed. Wild seedlings face risks from overgrazing.

The task is to encourage farmers and herders from passive to active agroforestry in their direct surroundings, from users of self-planted trees to tree farmers. It is the equivalent of the transition from gathering wild cereals, to planting them. An agroforestry revolution is needed to follow the agricultural revolution. Quite simply, the neolithic revolution must happen for trees.

This will require a change in attitude. Classical forestry activities aim at the distribution of trees to the rural population assuming that the more trees that are planted, the better the programme. It is not the number of trees distributed but their purpose that determines the success of social forestry. The needs of the rural population determine the required number of species and appropriate management system.

The common strategy for forestry development is to extend and decentralize the existing classical organization and infrastructure. Staff, nurseries, transport facilities are simply moved out of the forest into the rural areas. The idea is that more seedlings can be delivered to more farmers, in more regions, by moving more foresters closer to the farmers. Time and again, this form of intervention fails because the project definition assumes the forestry department is the focus for development not trees for local people.

The way forestry policies and projects are conceived and implemented contributes to many failures. Often, they are based on preconceived ideas about what the local problems are and how they should be solved. Planning takes place without involvement of the people concerned. Implementation of plans is taken to be a technical action without consideration of the experiences and local knowledge of the population or of their cultural and social attitudes towards trees. Management of environmental resources is only possible when local people are involved, because local people make their living from these resources. Too often antisocial forestry is the result. The planting of so-called economic trees such as eucalyptus plantations, on land formerly used by the local population, is one recurring example.

Social foresters work from a different premise recognizing that rural forestry problems can never be solved without the direct involvement and participation of rural people. They understand that afforestation programmes based on free seedling distribution have limited impact as they only serve as a kind of environmental welfare programme. Such programmes do not address issues that are critical to the maintenance of local people's production and consumption patterns, to the sustainability of livelihood systems. A reorientation of rural forestry efforts towards the active involvement of rural people by, for example, encouraging farmers to raise their own seedlings, is more effective than an increase in size of the local forestry organization or the number of seedlings distributed to farmers.

By looking at farm households in total context of existing land use practice a different picture emerges. Rural people have never seen forestry or tree planting as a separate activity from the rest of their production system. The management of trees and shrubs is part and parcel of their land use strategy. The indigenous population has a wealth of knowledge about its own needs and how to go about providing for them. It needs tree products, not trees. Indigenous management systems are not just a set of techniques. They are an integral part of people's lives, embedded in their world view. It is an entire agriculture. Box 3 is an example of such indigenous knowledge in practice in Tanzania.

Exploring what people know

For a long time it has been assumed that rural people exploit their near surroundings to get as much as possible out of them, without considering the sustainability of the land. In reality, however, existing management practices are often subject to many more or less formal rules, regulations, customs and traditions. This makes the control over, responsibility for, and accessibility to, woody vegetation different for different strata of the population. Relationships, including ethnic, class and gender relationships, usually determine the population's access to woody biomass.

Rural people possess not only knowledge about trees, shrubs and vegetation as a whole, but also about the ways to use that vegetation. More importantly they know about its management. They understand the reactions of vegetation to cutting, grazing or burning, the methods of regeneration after exploitation and the traditional rules and regulations to avoid over-exploitation. Their wealth of knowledge and its importance is illustrated in Box 4. The range of indigenous technical capacity is large. However, it is unrecognized by professionals who do not look for examples of local technical competence.

From area to area, there are important differences in the way rural people manage the woody biomass that exists in their environments. All these different management systems lead to a great variety in landscape. Often each village has its own way of managing hedges or planting trees, and rules and regulations to collect and harvest wood in the near surroundings. Landscape implies management: different landscapes imply different management systems.

Due to large-scale, external intervention, the management systems have been disturbed. National legislation, taking all trees outside private land to the State, and the demand for charcoal in the cities are examples of interventions. Bringing more land under agriculture is the most important intervention. Despite these

---

Box 3 Ecofarming in Kilimanjaro

The indigenous land use systems that evolved from extensive shifting cultivation to forms of sedentary farming, and which support a high population density (up to 400 people/km^2) while maintaining soil fertility, are of particular interest for development of ecofarming in the tropics.

The Wachagga way of farming on the slopes of Mount Kilimanjaro in Tanzania is a prime example (Fernandes et al. 1984). The forest, like farms, has several storeys of cultivated plants. Grevillia trees more than 20m high provide timber and fuelwood and improve the soil with their leaf fall. Beneath them are fruit trees like mango or avocado. The thinner trunks of the yang trees support twining yam plants. At a lower level a latticed canopy of leaves is formed by banana plants 4–5m high. Underneath the bananas are coffee shrubs 2m high, the main cash crop. Beneath these grow the shade crop cocoyanis (taro). The tubers are important to the Wachagga diet and the leaves serve as vegetables and fodder. Wherever more light reaches the soil, maize and beans are sown. Sweet potatoes and local vegetables are scattered here and there.

This multi-storey structure imitates the virgin forest which once covered the slopes. Weeds and pests are suppressed. The layer of organic matter covering the soil guarantees humus formation and to a large extent prevents erosion. Tillage is minimal, only planting holes are made. Cattle and pigs are stall-kept. Manure is collected and used as fertilizer, especially for banana plants.

This is one of numerous examples of highland farming systems that once existed in many parts of eastern Africa, from the Aberdare Mountains in Kenya to the Usambara Mountains in Tanzania. In rainforest and savanna areas too, productive systems of permanent land use were developed by local farmers. All these systems have certain elements in common. Many trees are scattered over the fields, species deliberately chosen to promote the growth of field crops. Organic matter in the form of mulch, compost and animal manure is carefully applied to maintain soil fertility. Livestock are stabled and grazing is limited to the period after harvest. In other words, all these systems have an agro-silvopastoral eco-design in which field crops, trees and livestock are closely integrated.

developments, it is important to explore what people know, not least because people preserve their knowledge in and around their homesteads. Hedges mark houses, fruit trees are found on compounds and special trees (for example, producing flowers for the women) are brought from outside as experiments near houses. Local landscape is an experiment in action.

Knowledge forms the base from which landscapes are built. The level of intensity is determined by other factors such as the land tenure system, the land use history and the physiological conditions in the area. The land tenure system, the degree of control or ownership over land and tree resources, in combination with the degree of primary management responsibility, enables us to distinguish various directions

---

Box 4 Indigenous knowledge

Traditional farming systems are classical low-external-input systems, making use of locally available energy and materials plus the practical environmental knowledge accumulated by generations of farmers. Communities with long experience in a particular area have developed techniques and strategies of resource use to suit the prevailing conditions. The very fact that their production systems have survived is a testimony to their ecological success. The mismanagement of the natural resources would have destroyed the basis of their livelihood.

The traditional techniques and strategies of subsistence farmers can be regarded as the result of 'deliberate efforts to improve and/or protect the value of life-supporting resources and to insure some reasonably secure long term viability' (Beyer, 1980)

Seen in this way, millions of small farmers throughout the developing world are agroforestry or ecofarming experts. It is important that indigenous knowledge be recognized and built upon by foresters for three reasons:

- Present forms of resource use which have long sustained the life of large numbers of people in resource poor or fragile environments must be preserved until provably superior forms of resource use have been developed
- Indigenous farming practices and environmental knowledge offer starting points for developing agroforestry or ecofarming measures which will increase the productivity and sustainability of local resources in developing countries, and
- Indigenous agroforestry knowledge can reveal missing ecological keys which may help foresters to develop alternative farming techniques in industrialized countries which will be less dependent on non-renewable resources than present systems and technologies of modern farming.

The small farmer's expertise represents the single largest knowledge resource not yet mobilized in the development enterprise. In order to develop sustainable agroforestry measures, foresters must learn from the local farmers to recognize and understand indigenous agroforestry systems and technologies so that these can be promoted, developed and diffused to other farmers operating under similar conditions.

---

for social forestry strategies to be found in the rural areas. A colleague, K. F. Wiersum, on the basis of these two factors describes nine different social forestry strategies. Three quite different kinds of management strategy emerge. The forestry management strategy is the least discussed but offers as much potential as the community on state management strategy. The strategies emerge from consideration of the tenure system and management systems outlined below.

Control/ownership of land and tree resources

		Community	Farm	State
Management	Community	1	2	3
of tree	Farm	4	5	6
and land	State	7	8	9
resources				

Social forestry management strategy characteristics

Community or communal forestry	1	Communal tree growing on community lands
	2	Tree growing on private lands organized by community institutions
	3	Public land allocated for community forestry projects
Farm forestry	4	Private tree growing on communal lands
	5	Privately managed tree farming plantations around households
	6	Public land allocation schemes for private tree growing
Publicly managed forestry for local community development	7	Public plantations on communal land.
	8	Public plantations on private lands, and
	9	Publicly managed schemes on public lands aimed at benefiting local people.

Exploring how people are

The word community is of little help when dealing with the heterogeneous nature of the social groupings. There is need for a concept that clearly acknowledges the heterogeneous nature of rural society and which can deal with this heterogeneity. This observation is as true of the developed world as it is true of developing countries.

The notion of an interest group is far more helpful. The concept refers to a group of people who have similar sets of interests in a particular situation. For example, people who own large numbers of livestock which are grazed on a patch of common land have different interests from people who have only a few stall-fed animals. A proposal to establish a plantation on common grazing land will affect each group differently.

Identification of various interest groups is fundamental to any project activity. The number of separate interest groups will differ according to different situations. A minimum list of interest groups would include:

- Women
- Landless
- The poor
- Ethnic minorities
- Religious minorities, and
- Specialist livelihood system interest groups.

The interest groups are largely self-explanatory with the exception of the last group. Examples of the last group are livestock owners dependent on common grazing land; blacksmiths dependent on forests for production of charcoal; and tea shop owners with a heavy demand for fuelwood. It may be necessary to make subdivisions in the specialist livelihood system groups.

The concept of interest groups is a tool which assists in the identification of relevant social groups in a heterogeneous society. It can also act as the basis of a checklist which ensures that all interested groups are involved in negotiations.

Women are the major interest group. The recognition that men and women are different social groups and may have different forestry interests is essential, and extensively documented. As forest and tree resources become more scarce, the balance between what people need and what they can obtain shifts. The effect on women is particularly severe because they depend on tree products and assume an ever-growing share of family work as men seek cash income by wage migration. Failing to recognize the importance of forest resources to women can lead to the introduction of technologies that cut off women from this critical resource.

Because women are the traditional users of trees and traditional systems of management are breaking down, women directly experience burden of disappearing resources. There are three basic negative impacts on women's lives. These are:

- Because traditional production systems are breaking down and trees are becoming more scarce, women have longer to travel to collect firewood or fodder
- As men migrate to find paid income, often to towns, this results in a labour deficit on the farm which women have to fill. This leaves not only less time for collecting firewood or fodder, but also less time for farming. Increasingly, lower labour inputs mean a decline in yield with consequent lower calorie intake. Women in rural areas frequently are the focus for the feminization of poverty as they lose control of resources within and beyond the farm, and
- Technologies and external influences are altering land use systems thus reducing the availability of forest products that women may use as a form of income.

Participation of women in social agroforestry projects is crucial to their success. This is easier said than done as many factors militate against women's participation. A key issue is land tenure. Because trees grow slowly, few farmers are prepared to plant trees unless they are sure that they will enjoy the benefits. They need security of tenure of land and trees. This is often a problem for men and almost always for women. Women without legal rights to land have no collateral to offer for loans to buy the equipment, seeds or fertilizer they may need. Women with no rights to use certain trees, as is common in many societies, have no incentive to plant them. Women who are forbidden by custom to plant trees have little chance to participate in forestry projects.

To be able to participate, women must also have the time, but they rarely do. The more women are likely to benefit from a project, the less they are likely to have time for it. This seems to be a general rule in our experience. For example, collecting fodder has become so time consuming for women in Nepalese hill villages that they are too busy to plant fodder trees.

In many societies, women do not enjoy the same freedom to travel as men, are not allowed to work away from home, or suffer from equally restrictive traditions. These are the areas where projects or intervention plans often fail because they are expecting activity from the women that is beyond the regulations of tradition. For example, in Kenya, women refused to join a honey-producing project because the hives were in trees because tree climbing is taboo for women. Lowering the hives solved the problem. In Sudan, the mobility constraint was overcome by moving nurseries into women's compounds. Problems of communication between male project staff and women can be solved by ensuring that women staff are hired. While not all constraints to women's participation are as easily lifted, policy measures can go a long way to help.

Women must be specifically, though rarely exclusively, targeted during project formulation. The fundamental need is to evaluate the potential impact on, and expected benefits to, men and women separately. Gender issues need careful analysis to avoid unintended effects on either sex. For example, a common mistake has been to introduce new crops or products that require heat processing or drying. These innovations may well increase men's incomes, but only at the expense of making more work for the women who collect the extra firewood. Enquiries need to be made into the needs, interests, talents and desires for participation of women in communities to be affected by forestry projects. Involving women in project design, as well as project execution, can eliminate some of the less desirable practices that occurred. For example, projects planned to employ large numbers of women in nurseries, because they could be paid less than men.

Another basic issue that needs further analysis is the role of women in the cash economy. Because women traditionally work in the subsistence sector, it is tempting to design projects to assist them in their traditional roles. In fact, women urgently need to be involved more fully in the cash economy. To achieve this, they should be provided with credit and security of land tenure on an equal basis with men.

Understanding the role of women is, however, difficult. In the southern part of the Peruvian Andes, temporary migration is a common phenomenon. More than fifty per cent of the men leave their farms for some months of the year to gain an additional income in the big towns. The women are left in charge of the whole farm. Many extension workers (almost all were men) in the Peruvian community forestry project concentrated their activities and communication on the men and complained about the bad participation in the maintenance of the community nursery in certain parts of the year. Initially, the idea of working with women did not occur to them. When, during several training sessions, the role of women in the community became clearer to them, many of them saw the necessity of co-operating with women. However, there remained a problem: although there is no taboo of men working together with women, many of the extension workers were unfamiliar with the operation of women's groups particularly mothers' clubs.

One community, in the Cusco department of the Peruvian Andes, possessed a large quantity of *Sambucus peruviana*. The mothers' club decided to start a production unit for marmalade from the fruits of this tree. Technical assistance (marmalade preparation, procedure and marketing) was provided by the community forestry project. The project lent some money for necessary inputs such as containers for the marmalade and sugar. By the end of the first year, the women were able to pay the money back, and buy containers for the next year. There was enough money left to be used for other items of expenditure selected by the community as priority investment such as disinfection of the houses. Besides raising income this activity raised the social position of the women within the community. Because of the success, men also wanted to establish their own small-scale forestry-based industry. They wanted to concentrate on the production of utensils made of the wood of *Alnus jorullensis*. In other communities the people became eager to plant more *Sambucus peruviana*.

Such examples are not restricted to Peru. During a cultural survey in Kenya, women indicated that trees with straight stems, like eucalyptus and black wattle (*Acacia mearnsii*), are typically the responsibility of men. This was because such stems made good building poles and house building was a male responsibility. They were also readily marketable in bulk and men control the bulk market. It is, therefore, difficult for women to harvest these trees as they are linked to land tenure issues and to the marketing of farm products. Only crops with limited seasonality, like bananas, could be planted by the women. Growing crooked trees opened up social forestry possibilities.

Throughout Africa, these ownership and rights problems exist. Certain interest groups share mutually recognized claims to specific user-rights of particular trees. Rights to use specific forest products also exist without the person necessarily being an owner of the land.

Society, history, custom and practice are all heavily influenced by culture. It is culture that sets limits, gives identity and gives a heterogeneous social system. This sets the bounds for management practices, intervention and action. For co-operation and mutual understanding, confidence in the external agent by the community is a prerequisite. The social forester needs to speak the local language and should be interested in the cultural aspects of the community with which he or she has to deal. To respect people and greet them in the proper way is more important than knowing the details of nursery technique. Feasts and religious events should be respected, and, if possible, the social forester should be actively involved in them.

Religious beliefs and trees are often intertwined. Fairy-tales, local traditional stories and myths may contain elements related to trees. They serve as possible starting points for discussions about the problems or benefits of trees.

The cultural gaps that exist between the external agent and the farmer cannot be bridged completely. These gaps may be narrowed if the outsider is prepared to participate in locally important expressions of culture — for example, songs, dances, births and weddings. Participation in such events is both a way of listening and a way of establishing the legitimacy of the social forester's presence.

How people are approached requires knowledge of the way they are organized. A project might visit every peasant separately, and try to convince him or her to co-operate with the project. If the project can, however, co-operate with an already

21

existing organizational structure within the village, the work may become easier. In some situations, it is simply impossible to co-operate with individual farmers without paying sufficient respect to the local organization. This, however, does not imply that, if a communal organization exists, the only possibility is to co-operate with the community as a whole. Not everybody in the community may be interested in the establishment of a communal nursery. In such cases, it may be preferable to work with a group of interested persons.

This emphasis on asking questions with people, exploring how people are, is not simply a different technical starting point. It is an approach that requires social foresters to conduct themselves in a different way. It is destabilizing because it asks more questions of the social foresters themselves than it asks of the community. It takes courage.

3. The Classical and the Social Forester

New and old knowledge: traditional versus modern approaches

Sciences do not operate by a neutral process of observation, hypothesis and empirical testing. They have distinct professional cultures, or sets of shared and implicit assumptions, that are rarely questioned. This implicit body of belief can also be called a paradigm.

The traditional forestry paradigm sees forestry as the study of trees and forests. All forestry knowledge is stored with the forestry. The forests are the central focus. Everything else, including people, is peripheral. The consequences of the prevailing forestry paradigm include:

- The reluctance of foresters to have the decisions concerning forestry really taken by the villagers, because foresters see themselves as having a monopoly of knowledge about forests
- The tendency to see other disciplines involved in community forestry (especially social sciences) in a service role, and
- A strong concern with detailed information about forests (full surveys and inventories) when such information is not needed for a tree planting programme.

Many foresters, although working in community forestry projects, still adhere to this traditional paradigm. However, during the past few years there has been a slow but perceptible change towards an entirely different paradigm. This new paradigm puts people at the centre of the forestry problem. Everything else is peripheral.

The basic changes in orientation of forestry can be summarized as follows:

From:	To:
Working for the interest of the forest department	Working for the interest of rural people
Forest management	Forestry extension work
Protecting forest against rural people	Involving people in management of woody biomass in and outside gazetted forest areas
Seedling production and distribution	Facilitating local tree regeneration
Plantation management	Total woody biomass management
Timber and pulp trees	Multi-purpose trees and shrubs
Standard forestry management systems	Systems that build on locally existing knowledge of tree and forest management

In the following sections we will try to specify what it means to put people at the centre of the forestry problem.

The changing role of the forester in forestry

During the 1970s and 1980s, opinion about the role and responsibility of forestry in the context of rural development has changed drastically. The work of the so-called classical forester is traditionally in the forest, usually State-owned, where he or she is responsible for the management and protection of natural forest or forest plantations. The sites are usually remote, frequently on more marginal land. If the forester is involved in production, or reforestation, the programmes usually require heavy machinery for the outplanting of single species.

The classical forester essentially has two roles as:

• A policeman of natural resources, and
• As an engineer of mono-cropped plantations.

Thus the classical forester is caught in two roles that make him or her seem anti people and anti environment respectively. The forester is seen as the fat controller.

Nowadays, with increasing awareness of the importance of trees for rural development, something else is expected from foresters. They are expected to work for, and together with, rural people. They are expected to work with, not against, local land use systems. This kind of work has little in common with the job foresters have been doing to date. It implies that foresters have to shift their attention from a situation where they were almost in full control to one in which they control very little. In short, they have to change from being forest managers to rural planting advisers.

In order to clarify what this shift in orientation implies, it is useful to have a closer look at the two types of foresters. The first is a classical forester who is only marginally involved in rural afforestation. The second is a social forester who has been trained in rural afforestation (or social forestry) and has working experience in that sector.

Box 5 typifies both foresters and their contrasting approaches. The questions a classical forester would ask, even when dealing with rural forestry, would emphasize physical supply and technical parameters. In contrast, the social forester would emphasize demand for forest products, current land use practice and the role of local farmers. This division between classical and social foresters is caricature. Real life is a continuum, a sliding scale with few people on either end but many somewhere in the middle.

It is obvious that foresters have a certain knowledge about, and experience with, trees. It is often automatically assumed that foresters should be able to advise farmers and non-forestry development workers about trees, how to grow them and how to manage them. But foresters frequently know very little about the priorities and needs of rural people. The knowledge foresters possess is largely irrelevant because it has been developed in the context of plantation forestry. Plantation forestry has different goals, uses different species and employs different techniques from social forestry.

The belief that forestry techniques for rural development are no different from the forestry techniques used in plantation forestry is still widespread. Tree nurseries, planting methods, spacing, rotations and harvesting methods are assumed to be the

---

Box 5 Forestry questions

Ten questions of a conventional forester

1. How many hectares of forest do you have under your supervision (natural, plantations)? How many hectares did you harvest? How much is cut illegally? How many hectares did you plant?
2. How many nurseries do you have under your supervision?
3. How many seedlings do you raise every year?
4. What species do you raise? What do you think of eucalyptus; do you have an alternative?
5. Do you have problems with nursery inputs (supply, time or arrival) for example, polythene bags, seed, water, soil? How many nursery staff do you have?
6. How many seedlings have you distributed/sold to farmers; do you have logistical/transport problems?
7. How do you encourage farmers to plant more trees, to collect or buy more seedlings from the nurseries?
8. Do farmers use the correct method of tree planting and spacing?
9. How do you expect to cope with the increasing demand for seedlings?
10. How many years do you need to cover this whole area?

Ten questions of a social forester

1. How large is your working area?
2. What area is under woody vegetation: natural forest, plantations, trees on farms etc.?
3. What products do the people get out of the forest? Who has access to the forest and what is the area of influence?
4. What types of woody vegetation are found outside the forest? What is the pattern and composition of this vegetation in different zones?
5. What are the different functions of woody vegetation on the farm (productive and environmental)? What are the farmers' priorities?
6. What role does the farmer play in establishment, maintenance and management of woody vegetation?
7. What is the present role and priorities of the forest department?
8. What are the major constraints for development of the on-farm woody biomass; on the side of the farmer, the forest institutions, other institutions, physical and so on?
9. What could the forest department/project do to remove these constraints?
10. What could the role of the farmer be?

---

same. For example, it is assumed that timber trees, planted in plantations, could be developed easily by farmers. Yet surely it is common sense that a completely new set of forestry methods and techniques should be developed for rural development where the purpose and scale of planting is so different. Common sense, however, appears to be a rare commodity in rural development. Common sense techniques for social forestry would be small-scale, multi-purpose, flexible and based on the existing knowledge and experiences of rural people, emphasizing a wide range of appropriate species that give a wide range of products.

In Box 6 a comparison is made between some standard forestry techniques, as they are commonly applied in forest establishment and management, and techniques that are more relevant to rural tree planting or farm forestry. The classical forester prefers a more intensive, more managerial, more expensive approach: the goal is standardization. The social forester starts from the other end, working with non-standardized resources for the best result available at least-cost.

The objectives of classical forestry activities are the distribution of tree seedlings to the rural population. It assumes that the more seedlings that are planted, the better the programme. But it is not the number of distributed seedlings but the actual end use of the trees that determines the success of social forestry. The needs of the rural people determine both the required number of species and the appropriate management system. However, tree management strategies in rural areas have received little attention from foresters to date.

The common strategy for rural forestry development is to extend and decentralize the existing classical organization and infrastructure. Staff, nurseries and transport facilities are simply moved out of the forest into the rural areas. The idea is that more seedlings can be delivered to more farmers, in more regions, by moving more foresters closer to the farmers.

But rural forestry problems can never be solved without the direct involvement and participation of rural people. Concurrent with this is the fact that afforestation programmes, based on free seedling distribution, often have a limited impact. This is because they only serve as a kind of environmental welfare programme and do not address the issues that are critical to the maintenance of local people's production and consumption needs, or to the sustainability of livelihood systems. A reorientation of rural forestry efforts towards the active involvement of rural people by, for example, encouraging farmers to raise their own seedlings, is more effective than an increase in size of the local organization or the number of seedlings distributed to farmers.

Classical foresters frequently treat trees as an independent component of the farm, to be planted in separate blocks (woodlots) on second choice, marginal sites which are not productive for crops. To avoid problems with individual farmers and professional agriculturists, they recommend that trees are planted away from crops or livestock, thus separating the trees from the farmers. This has the effect of marginalizing the farmer who comes to see the tree as separate from his or her living reality. In contrast, social foresters see trees as an integrated part of the farming system both in ecological terms (soil conservation and soil quality enhancement) and economic terms (income generation). Tree, crop and livestock development cannot and should not be seen in isolation from each other. Tree development can support other farm activities and vice versa.

This caricature might be harsh, but it contains disturbing tendencies. The classical forester arrives in rural forestry programmes assuming a supply-side technology. In contrast, the social forester assumes nothing but seeks to evolve a programme which is demand-driven. The classical forester has a solution in search of a problem. The social forester has a problem in search of a solution. Box 7 lays out the characteristics of the different types of foresters by plantation or rural focus.

Box 6 Characteristics of forestry techniques on platations and social forestry

CLASSICAL FORESTRY

Nursery techniques
- Certified seeds are used
- Seedlings raised in polythene tubes, rarely bare rooted
- Special soil mixture
- Regular irrigation
- Fertilizer
- Pesticides
- Artificial shading
- Root pruning
- Labour-intensive

Planting methods
- Standard seedling size
- In big holes
- Fertilizer or manure
- Watering
- Replanting (beating up)
- Aim: high survival

Other regeneration methods
- Direct sowing (rarely)

Management
- Monoculture
- Regular spacing
- Regular weeding
- Regular thinning
- Regular pruning
- Fixed rotation
- Management geared towards production of uniform product: timber, poles, pulpwood, and
- Management rigid.

SOCIAL FORESTRY

- Seeds are locally collected
- Seedlings usually raised bare rooted but sometimes in locally made containers (milk-packs, old tins, banana fibres)
- Locally available, fertile soil
- Irrigation not necessary
- Farmyard manure
- Ashes against insects
- Under bananas or trees
- No root pruning but early transplanting
- Labour-extensive

- Varying seedling size
- In small holes, made with a pointed stick
- No manure
- No watering: plant on a rainy day
- No replanting, and
- Low survival is accepted

- Direct sowing
- Collecting wildings
- Planting cuttings

- Often mixed with other tree species and crops
- Varying spacing, often very close
- Little weeding
- Thinning if small sticks are required
- Pruning of branches are needed, or shade reduced
- Variable rotation, depending on required products
- Management geared towards a range of products: twigs, sticks, firewood, poles, fence posts, timber
- Management flexible, and
- Other management methods: pollarding, lopping, root pruning, ringbarking.

Foresters can only play the role of the social forester if they are willing to come out of the forest and listen to, and work with, rural people and other rural development organizations. This does not imply that social forestry should or could replace plantation forestry completely. There will always be a place for forest plantations. But, it is increasingly clear that, in many rural areas, forestry activities have sometimes been counterproductive to developing self-sustainable rural forestry

---

Box 7 Characteristics of forestry techniques in traditional and social forestry

Traditional foresters

1. Foresters are trained and have knowledge and experience with trees. It is automatically assumed that foresters are able to advise farmers how to grow, plant, manage and harvest trees.
2. Forestry techniques for rural development do not have to differ from standard plantation forestry techniques, for example, tree nurseries, planting methods, spacing, rotations, harvesting methods. Timber planted in plantations, should be planted by farmers as well.
3. Many forestry activities aim at the distribution of tree seedlings to the rural population. A common view is that the more seedlings are distributed the greater the success of the project.
4. A common strategy for rural development support is to extend and decentralize the existing organization and infrastructure, staff, nurseries and transport facilities. All simply move out of the forest into the rural areas.
5. Trees are seen as an independent component of the farm, to be planted in separate blocks (woodlots) on second choice sites which are not useful for crops.

Social foresters

1. Social foresters accept they know very little about the priorities and needs of rural people and that their knowledge is not necessarily relevant to tree growing on farms.
2. Social foresters argue that new forestry methods and techniques should be developed for rural development. These should be small scale, including multi-purpose species, and based on rural peoples' own practices. Timber is only one of the very many tree products required by rural people.
3. Social foresters argue that it is not the number of distributed seedlings but the actual end use of the trees that determines the success of rural forestry activities.
4. Social afforestation programmes, based on free seedling distribution, often have a limited impact. Active involvement of rural people, for example, by encouraging farmers to raise their own seedlings, is more effective than increasing the number of seedlings distributed to farmers.
5. Social foresters argue that trees are an integrated part of the farming system, both in ecological terms (soil conservation, improving micro-climate) and economic terms (income generation).

Box 8 Eight statements about rural people and tree planting

Traditional foresters	Social foresters
1. Rural people have little or no knowledge of tree planting.	1. Rural people have much knowledge about the utilization and management of trees and shrubs.
2. Rural people tend to exploit the natural vegetation beyond the limits of sustainability.	2. Rural people are able to manage woody biomass in a sustainable manner but are often forced, by factors beyond their control to over-exploit the resource.
3. Efforts should be made to educate rural people about the importance of trees and shrubs for their daily lives and the environment.	3. Rural people are clearly aware of the importance of shrubs in the local environment.
4. Since rural people lack adequate knowledge, they should be taught and instructed how to plant and manage trees.	4. Rural people need encouragement and technical assistance to enable them to do what they think is best.
5. If properly motivated, rural people will be willing to plant trees.	5. Planting trees is just one of the possibilities of woody biomass management; people might have very good reasons not to plant trees.
6. If they can be motivated, and the right incentives are given, rural people can participate effectively in tree planting programmes.	6. Rural people should be assisted to identify and design the most suitable tree management activities themselves.
7. Rural people tend to look at woody vegetation as a component that competes with other agricultural land use and, therefore, woody vegetation should be separated as much as possible from other crops.	7. Rural people, in principle, consider trees to be an integrated part of land use. Modern agricultural practices ignore the role of trees and thus cause the complete separation of trees from the rest of the agricultural land use system.
8. If people plant trees, the benefits will be distributed in the community according to needs.	8. The issue of who will benefit from tree planting depends very much on the ownership, user rights and control over woody vegetation.

development. Woody vegetation that is present in most rural areas is, in most cases, almost entirely the result of rural people's own initiative to regenerate woody biomass.

Social forestry can contribute by filling the gap between the techniques that plantation forestry has to offer and what rural people already know. In other words, social forestry must bring these two knowledge systems together, merging them into something that is better. The challenge of social forestry is to seek change by managing the negotiation between these two knowledge systems.

Over the past few decades important changes have taken place. They are still taking place. Generally speaking, forestry activities have moved 'out of the forest' into rural areas. Many non-forestry organizations have started tree planting activities. This movement out of the forest requires a substantial reorientation of forestry interventions, which implies a move from managing things to facilitating local action.

The glasses through which the traditional and social forester look at rural people are completely different. The traditional forester considers himself or herself to know much more about forestry than the local people. The social forester has a completely different attitude: he or she believes that rural people may know a great deal about forests, trees and possibilities to make them useful. Therefore, the social forester is prepared to learn from the rural people and to search with rural people for solutions to problems. Some other differences in attitude toward rural people are summarized in Box 8.

This chapter has shown three levels of change in forestry that are on-going because foresters have taken up the challenge of working outside the forest with rural people:

- Foresters themselves are addressing new sets of questions as 'social' rather than 'traditional' foresters
- These questions have led to a reorientation of work with an emphasis on accurately defining the problem rather than delivering a solution, and
- This has led to the reorganization of service delivery.

There is certainly movement. To judge whether it is movement in the right direction, it is necessary to see what rural people can offer the forester.

4. The Way Forward

Understanding existing management: the least cost solution

For a long time, it has been assumed that rural people exploit forests and woodlands to get as much as possible out of them, but do not manage the vegetation in a sustainable manner. This assumption accepts that people use woody biomass in much the same way that hunters and gatherers used the whole environment before the neolithic revolution led to the domestication of plants and animals and, consequently, the creation of permanent settlement.

Local management practices are often subject to many, more or less, formal rules, regulations, customs and traditions. This makes the control over, responsibility for, and accessibility to woody vegetation different for different interest groups in the population.

Rural people possess not only knowledge about trees, shrubs and the vegetation as a whole. They know about ways to use that vegetation. More importantly, they know about its management. They understand the reactions of vegetation to cutting, grazing or burning, the methods of regeneration after exploitation and the traditional rules and regulations to avoid over-exploitation. The range of technical capacity is large. However, sadly, it is unrecognized by professionals who do not look for examples of local technical competence.

From area to area, there are important differences in the way rural people manage the woody biomass that exists in their environment. These different management systems can be grouped into three broad categories namely:

- Tree management systems based on natural vegetation
- Tree management systems based on natural and planted vegetation, and
- Tree management systems based on planted vegetation.

Tree management systems based on natural vegetation

In many sparsely populated areas, woody biomass management consists largely of restricting the harvesting of wood from natural vegetation. In these situations, where people depend on the natural forest for a wide range of products, they have learnt which species are useful for what purposes. The cutting of trees is rarely indiscriminate. People have learnt to obtain most out of the forest without harming its productivity.

Interventions like cutting and clearing, temporary use as farm land, grazing, burning, harvesting of fruits, bark, branches, leaves and roots are often subject to rules or regulations which avoid exploitation beyond the carrying capacity of the vegetation. Specific measures are taken to enable the vegetation to restore itself, such as burning, complete exclusion of fire, rotational grazing and prohibition of cutting of certain species.

The slash and burn, shifting cultivation or extensive livestock production systems provide examples of systems that were both efficient and stable. The viability of these systems, however, is being eroded by a combination of external and internal pressures that destroy traditional sustainable practice without offering local alternatives for people to maintain their livelihood. Even so, the knowledge and experience on which these systems are based is still a valuable resource, perhaps as valuable as the genetic resource base itself. Intensification of production in these systems is necessary by developing woody biomass systems, either based on planted trees and shrubs or improved management of the natural vegetation.

In Box 9 an example is given of woody biomass management in dryland areas in Zambia. The *Chitemene* system is practised under increasing pressure as a result of population growth. Even though the system may be outmoded, because many external changes have led to a different pattern of local ownership and land use, the knowledge and experience on which they are based is valuable.

Management systems based on natural and planted vegetation

In this category, there are situations where farmers rely on natural woodlands for a substantial part of their requirements. Planted trees and shrubs play an important role as well. For example, there are fruit, shade and ornamental trees around

---

Box 9 Tree management systems based on natural vegetation

The *Chitmene* System, Zambia

The *Chitemene* system of slash and burn cultivation is practised in the Northern and Luapula Provinces of Zambia. It is an example of a successful tree management system in a dry area, that involves no tree planting at all but is entirely based on the regeneration capacity of the natural vegetation.

A farmer moves into an area and clears a patch of bush. Additional trees are also cut from the surrounding areas and carried in. These are all then burnt and, in the clearing that is left, a garden is established. The neighbouring bush is left to regenerate.

Ashes from the fire enrich the soil. Recent evidence suggests, however, that it is the heat from the fire that may be even more important in encouraging crop growth as it alters the balance of micro-organisms in the soil in favour of species that are able to fix nitrogen. Timing of planting is critical. Farmers know that it is best to wait some time before planting, allowing nitrogen fixation to begin, but not too long as weed encroachment will become a problem.

The *Chitemene* system has served farmers well for generations. Gardens are abandoned after a few seasons and farmers move on to clear new areas. Provided the land is left to recover, the system is entirely sustainable. The bush recolonizes the area and the natural balance of nutrients and vegetation is restored. Traditionally the fallow periods were at least 25 years. Nowadays, the fallow periods have fallen to only 10 years in places as a result of the expanding population and the reduction of available area. The long-term impact of these changes is still unclear, but the likelihood is that crop yields will gradually decline.

Box 10 Multi-purpose farm woodlots

Small woodlots often provide more than poles and timber. Depending on the species that are planted, a wide range of products can be harvested from the same piece of land. Below is an example of a black wattle (*Acacia mearnsii*) woodlot in the Kenya Highlands.

Establishment
The woodlot is established by sowing black wattle seed directly into the cultivated land, either broadcast or in lines. The result is an extremely dense woodlot. The density has the advantage that closure is quick and weeds do not get a chance to develop.

First harvest
After one to two years a first thinning is possible. At this stage the trees are tall and flexible and can be used for the following purposes: firewood, sticks to support tomatoes, twines to make baskets or granaries, sticks to make the framework of the walls of huts or the horizontal layers of fences.

Second harvest
After two to four years, a second thinning can be carried out. The trees have grown to the size of poles (10 to 15cm diameter) and can be used as fence posts, for building (especially roofs) or for firewood.

Third harvest
At six to ten years the remaining trees have developed into big, strong trees that can be used for a variety of purposes: building posts; fence posts; and split wood for firewood or charcoal. The bark of the trees is sold to the tannin industry.

Looking at the composition of the woodlot during different phases of the complete rotation (see the graph below) it is obvious at any stage there are wood products to be harvested and that the type of product relates to the density of trees.

For eucalyptus, another species used for woodlots, the same applies. Some differences occur such as the fact that eucalyptus is planted instead of sown, is less suitable for making granaries and, if old enough, is suitable for sawn timber as well.

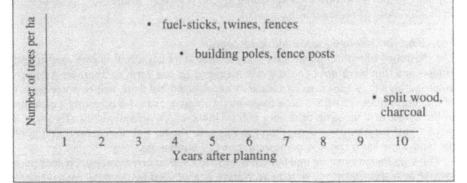

33

homesteads, hedges for demarcation or fencing and small woodlots for poles. In these situations there is often a wealth of knowledge about planting trees (both indigenous and exotic species). The knowledge about the indigenous vegetation, and the species that grow in the local environment, is considerable.

In many such situations, there is a tendency to clear the remaining natural vegetation and to move towards planting of few exotic species. In these situations, the combination of new tree planting experience and traditional tree experience can be fruitful, leading to stable and diverse production systems.

The trees that usually disappear first from natural forests are the straight ones, for poles or timber. Often, people respond to an increasing difficulty to obtain building material by establishing their own small woodlots, usually of fast growing exotic trees, such as eucalyptus or black wattle (*Acacia mearnsii*). This, again, is unsurprising because everyone needs to build their own home. In fact, a good method of measuring wood scarcity is to ask how many local farmers produce poles for building — a drive to self-reliance — rather than rising fuelwood prices. Apart from building materials, these woodlots often provide a number of other products as well. In Box 10, there is an example of multi-purpose woodlots of black wattle in the Kenya Highlands.

Tree management systems based on planted vegetation
In many parts of the world, in particular in densely populated, high-potential areas, where the indigenous forest has been cleared for some time, people rely entirely on their own tree planting activity. Intensive and complicated systems have evolved in which different species are found in many arrangements, such as mixtures of trees and crops, multi-storey home gardens, wind-breaks, woodlots, living fences and hedges.

Although much traditional knowledge about forest species has disappeared, local knowledge about use and cultivation of planted species (both indigenous and exotic species) is considerable. The scope for development of these systems is great, as woody biomass management is fully integrated into people's lives.

Good examples of these kinds of systems can be found in parts of the East African Highlands where most of the natural forests have been cleared to make room for agricultural land. Kisii is an area where farmers have managed to integrate tree planting into their farming system, in response to the disappearing natural forest (see Box 11).

The Kisii example, from Kenya, is not unique. Every farmer has an implicit or explicit system of woody biomass management. Nowhere is this more evident than in the diversity of regeneration practices.

Searching for managed trees and landscapes
The existence of natural landscapes is somewhat of an untruth. It is very rare to find landscapes that have not been actively managed in one form or another. Areas or environments may appear to be natural or unmanaged but there will be some aspect of that environment that has been consciously created, even if it is simply a clearing in the forest as a meeting or resting point. However, environments usually show a more significant human impact than the example above, and often every section of the landscape has a specific purpose in terms of human use.

The conclusion must be that human interaction and intervention has created (to a greater or lesser extent) environment. Nature is produced by the local management

34

Box 11 Kisii, Kenya

Kisii is an area in east Africa where farmers have taken up tree planting as a normal, almost routine, activity in response to the growing needs for tree products, such as building materials, fencing material, wood for furniture and tools, firewood and charcoal. The area has a population density of about 600 people/km², a high agricultural potential, high rainfall and fertile soils. All land is privatized, but there are few, small forest plantations. Natural forest has been almost cleared completely. In spite of the pressure on the land, about eighteen per cent of the area is under some kind of woody vegetation: woodlots, windbreaks, hedges and bushes. The different ways farmers manage woody vegetation on their farms emphasizes the range of different management systems. In Kisii there is both planted and natural vegetation. Planted vegetation includes woodlots, windrows or windbreaks, hedges, trees in and along cropland and trees around the homestead. Natural vegetation includes woodland and bushes.

Planted trees and shrubs are nearly always exotic species, such as *Eucalyptus spp*, cypress (*Cupressus lusitanica*), black wattle (*Acacia mearnsii*), mauritius Thorn (*Caesalpinia decapetata*), *Lantana camara* or *Grevillea robusta*. Few indigenous species are planted or protected although there are some *Sesbania sesban, Markhemia platycalyx, Croton macrostachys*. Other indigenous species are still found scattered around, but are quickly disappearing. Fruit trees are rare. Loquat (*Eriobotrya japonica*) and avocado (*Persea americana*) dominate.

There are four different ways in which trees and shrubs are regenerated. First, there is **planting of seedlings** (mostly eucalyptus and cypress). This is regularly done by about fifty per cent of all farmers. Most seedlings are raised by the farmers themselves in small on-farm nurseries. Occasionally they get them from neighbours, buy them from the Forest Department or on the market. Second, there is **direct sowing** mostly black wattle, *sesbania* and hedge shrubs. Seed is collected locally or bought on the market. Sowing can be done broadcast, in lines or in individual holes. Third, there is **planting of wildlings**, usually *Grevillea*. The emphasis is on transplanting or protecting naturally grown seedlings. Finally, there are **cuttings** especially of hedge species. Planting of branches or stem-sections are densely placed to get rapid closure.

Eucalyptus, black wattle and cypress are planted away from cropland in swampy areas, on hilltops, and on roadsides. Other species, *Grevillea, Sesbania, Croton, Markhamia* and some hedge shrubs, are planted along or in cropland (maize, beans, coffee, pyrethrum), grazing areas or on farm boundaries. Fruit trees are often planted near the homestead, on the compound or in nearby fields. Vegetable gardens are a favourite site.

In woodlots, spacing is very close to avoid weeding, to get an early closure and to get thin sticks and poles for small construction purposes. In cropland spacing is wide, depending on the tree species and crop. Weeding is rarely done for tree crops but those in cropland benefit from the weeding of the crop. Pruning and pollarding trees planted in, or near cropland, is regularly undertaken to reduce shading and to get firewood or branches for other purposes. Finally, coppicing of some species occurs to encourage regrowth after cutting, for example, eucalyptus, *Markhamia, Croton, Combretum*. Usually a few stems are left to grow into mature trees, while the rest is cut as sticks or poles.

of natural resources. This is a logical conclusion and a logical starting point for biomass intervention. If particular people depend upon an environment for subsistence, then they will manage it in the most sustainable and productive way possible. It is against their interests to damage it, because to destroy the environment would be to threaten their survival. It is largely when external influences invade indigenous systems that there is a breakdown in local management practices.

In terms of social forestry, because rural people have always lived with trees and shrubs they, not surprisingly, know about them. However, until recently, few foresters tapped this local knowledge about the complex role that trees play within local production systems. Information about tree species planted, or occurring in natural vegetation, is relatively scarce. Even less is known about why farmers prefer certain species and what methods they use to grow them. Most surprisingly, nobody puts the total picture together, that is that local people knowingly produce the landscape.

The question to be asked at this point is why have foresters ignored or not seen this phenomenon? Essentially, it is a problem of not seeing the trees. Woody biomass is often dispersed and varies in structure. Most foresters and agriculturalists do not look for it so it is not surprising they do not find it. Rural development professionals are strongly biased towards crops or livestock and pay little attention to trees and shrubs. Their perception of a built landscape is an orderly production system where each sector can be seen separately, that is trees in plantations, and crops in fields.

Foresters tend to overlook anything that does not look like a real tree, such as small trees, shrubs, hedges or bushes. Woody vegetation only appears in land use statistics in the form of forestry plantations or natural forest. Smaller elements of trees and shrubs on farmland or communal land are hardly mentioned. Yet, in many different land use systems where trees are assumed unimportant, the percentage of the land covered by some kind of woody vegetation may be somewhere in the range of ten to thirty per cent. The very existence of such large volumes of woody biomass vegetation clearly indicates the importance of trees and shrubs for rural people. This importance has manifested itself in management techniques that have evolved over time.

An important conclusion is that foresters have to look in another way at the various landscapes that they work in. The classical forester, and landscape architects, look first at their geological map, the soil map and topographic map before making an interpretation of the land they see. But these maps often do not carry the information they need. The social forester is interested in landscape elements and their relation to people. What do these landscapes tell us about the management practices in use? The social forester will look at hedges, woodlots, single trees and windbreaks wondering what management might exist behind the phenomenon he or she observes.

As a preliminary exercise the social forester will take a picture of a landscape in their working area and interpret it. Are there hedges? Are they prolific? What could they be used for? What species are used? Are there single trees? Are there trees around the houses? Just looking at the landscape, helps identify production systems.

Evolution of the landscape
The evolution of the landscape has taken place globally and not just in developing countries. To illustrate the importance of recognizing managed landscapes, consider the evolution of landscape in The Netherlands.

In farming areas to the north and east of The Netherlands, local traditions still exist to support the land use rights of local farmers. If they need poles to repair their houses or to build a fence, customary law allows the farmers to go to the forest to harvest a limited number of trees. Today these practices are not so common but they give an indication of the traditional relationship of the farmers to trees.

In the past, farmers needed access to the forests to construct houses, build fences, collect fuelwood and charcoal and harvest the by-products of the forest to supplement their farm income. After the introduction of fertilizers, the situation changed with the development of intensive farming systems which concentrated on the production of crops and livestock, leaving forestry to specialized government organizations. Forestry, and the production of trees, became a specialized profession with limited links to the agricultural food and livestock production systems.

In the farming areas, which became very intensive agricultural production systems, trees totally disappeared. The agricultural landscape in the west of Holland is covered with glasshouses, developed for flower production, or dairy farming. Trees on such productive, but often water-logged, land are only found in and around the farms or along the ditches and canals. The species are limited, for example, to willow (*Salix spp.*), alder (*Alnus glutinosa, A. incana*) and poplar (*Populus spp*).

In eastern Holland, with sandy and less water-logged soils, more trees are found on farms. Farming in these areas was difficult and much of the land was owned in large holdings which were reforested as hunting grounds. Large beech (*Fagus sylvatica*) and oak trees (*Quercus robur*), alternating with rhododendron, are typical remnants of this landscape building activity. When farming on the sandy soils became more economic, because of the use of fertilizers, small farmers started to occupy these lands as well. Tree planting continued but beech and oak were replaced by alder and poplar, a common sight in the rural landscape of The Netherlands.

Land too poor for intensive farming was exploited or left to be grazed by sheep. Large parts of the Veluwe, in the central part of The Netherlands, were covered by heather and sand dunes. Soil and wind erosion were common. Production on the land was deliberately low to halt erosion. To increase production, large areas were reforested by private and government organizations. Most of these man-made forests are today managed by the State Forest Service, but private organizations also play an important role in maintaining the forests.

This short overview of tree production in landscapes from The Netherlands has striking parallels with the present situation in many developing countries where similar changes are taking place. The original vegetation has been replaced by a built landscape.

A study in a Gambian rural village, Jappeni, Nicki Allsop showed how a tree management system, based on both indigenous and planted vegetation, has developed to ensure a diverse and stable biomass economy which meets the subsistence needs of the whole community. The villagers of Jappeni have in-depth knowledge of their own environment, the precise agro-ecological conditions and the multiple products and services which can be obtained from different tree and shrub species. This wealth of indigenous knowledge is reflected in the local system of land management which farmers have used to knowingly create their own landscape.

Figure 1, a sketch map of an individual compound and home garden, and Figure 2, a sketch map of a representative section of the local tree resource base, show that

37

the local tree resource base can be divided into two broad categories each with subdivisions, namely:

1. Trees and shrubs on areas of common property.

- Permanent grazing land/communal woodland
- Fallow cropland
- Open areas within the village itself, and

2. On-farm trees.

- Rice land
- Rotational cropland
- Home gardens.

The planted component, which is effectively limited to home gardens, has arisen as a rational response by individual farmers to certain constraints in the use of the indigenous woody resource base. These constraints reflect seasonal labour shortages, limitations on access and other culturally determined patterns of behaviour. By introducing the planted component, farmers are employing a successful risk management strategy. It helps to diversify farm production by supplementing and complementing outputs from the naturally occurring woody biomass vegetation, provides products, services and income needed during the period between main harvests and enables villagers to bridge the peaks and troughs in seasonal demands for labour.

In the south-western province of Kenedougou, in Burkina Faso, other workers observed a well-forested landscape that is threatened because of free grazing, burning and over-consumption of wood. But the cultivation of fruit trees is widespread and changes the landscape. The tree savanna in the south, with much gallery forests, changes gradually and becomes a landscape with a slash and burn fallow system with cultivation of yam, cereals and the plantation of fruit trees from the original vegetation are left, like karite (*Vitellaria pradoxa*) and nere (*Parkia biglobosa*). New vegetation is often more dense than the original. Trees are an integral part of this landscape. Trees, in small woodlots, along the ditches or in so-called wood walls, are not the responsibility of any specific organization; only forests are managed by the organizations. But management systems clearly exist for the trees outside the forest even if it is unknown.

In the province of Oubritenga, again in Burkina Faso, in the village of Nahatega, Joseph Tabsoba's farm is rich in trees. The field of half a hectare has clear boundaries of stems of neem (*Azadirachta indica*), interwoven with dead branches of *Zyziphus mauritania*. Cassava, pepper, eggplant (aubergine) and maize were cultivated. The provenance of the neem tree is from Ghana. The plantation is 12-years-old and already coppiced once. Within the yard, surrounded by the neem tree-zyziphus hedge, the following tree species were found: *Psidium guajabe, Magnifera indica, Tectona grandis, Terminalia avicennioides*. Spacing was 6-10m, so crops could still prosper. In the yard the farmer has his mini-nursery with approximately 80 seedlings, to catch the morning sun. The limits of the fields are demarcated by exotic tree species like neem, *Cassia siamea, Eucalyptus camadulensis*, which are also used as a wind-break. Anti-erosion bunds are made with stones reinforced by grass of *Andropogon gayanus* and trees of *Cassia siamea* and *Bauhinia frutescens*. Mulching is practised to conserve soil humidity. In the fields themselves, the farmer has planted *Parkia biglobosa* and *Eucalyptus camadulensis*. The latter suffers attacks by termites. Natural regeneration of local species is practised, but seems to be difficult. The farmer actively protects indigenous trees, integrating them into his production system. It is a conscious process of management that has resulted in a built landscape.

In developing countries, the management of trees left in and around agricultural land, or trees planted by farmers is unrecognized and underdeveloped. The on-going landscape building process is underestimated and the role of farmers in the process is unclear. Only the destructive acts of farmers are highlighted, not the constructive efforts they undertake. But, as in The Netherlands, the rural population in developing countries are actively fighting against land erosion. They are fighting to build a new landscape.

It is important, however, to offer the right tools for this effort to build a new landscape. The development of fast growing species, like alder and poplar in The Netherlands, was a landscape management tool. Today these species are common, but the memory of beech and oak landscapes lingers. Tradition frequently defines perception so that, even today, beech and oak are considered to be real trees in contrast to alder and poplar. Alder and poplar are known for their soft wood and limited use, but they grow three or four times faster than traditional beech oak combinations and are therefore economically attractive to farmers.

In rural areas in developing countries, original tree cover will not meet the demand for wood. The result is over-exploitation of natural forest which, combined with the demand for agriculture land, leads to the disappearance of indigenous forests. New landscapes appear which, in the beginning, have few trees. During the years following colonization, trees become more common. Particularly in areas with high population densities, man-made tree landscapes are very common. The home gardens of Java, Indonesia, the Kandy gardens in Sri Lanka and the farm trees in the Highlands of Kenya are interesting examples of new, man-made tree landscapes. This clearly suggests there is no simple correlation between overpopulation and deforestation.

It is often assumed that the amount of woody biomass decreases with increasing population density, or expressed in popular language, that overpopulation causes deforestation. This is not necessarily true. In many places, people respond to an increasing need for tree products by increasingly planting or protecting trees, either for their own use or for sale. Of course, there are many examples where increased population leads to a decrease in woody biomass availability. But, even where such a situation occurs, there is no reason to assume that the local population do not wish to reverse the trend. The desire to plant trees still exists. However, the support to do so — the external inputs — is frequently poor.

There are often clear patterns in changes in the woody vegetation. For example, the number of planted trees often increases with increasing population density, while simultaneously the amount of natural vegetation usually decreases. Again, the number of commercial trees often increases near urban areas where there is a market for poles or firewood. What is essentially occurring is that woody biomass is undergoing domestication in new settlement areas or areas much more populated than in former times.

This domestication can be compared with the neolithic revolution in which wild herbs and animals were domesticated to become agricultural crops and animals. This domestication was accompanied by the gradual change of mankind from hunter

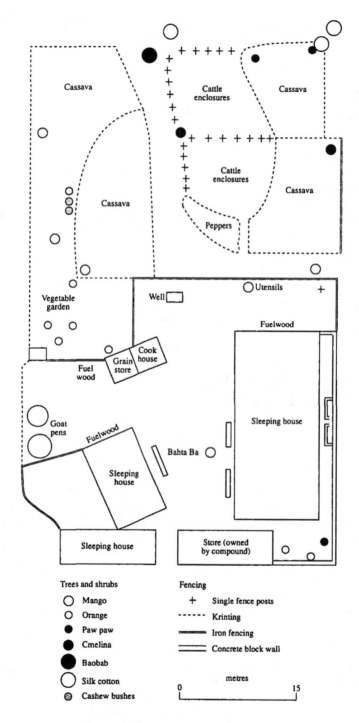

Trees and shrubs

○ Mango
○ Orange
● Paw paw
● Cmelina
● Baobab
○ Silk cotton
◉ Cashew bushes

Fencing

+ Single fence posts
----- Krinting
=== Iron fencing
≡ Concrete block wall

metres

0 _____ 15

Figure 1. Compound and home garden

and collector to settled farmer. To date, woody biomass has been a free good, not least because it is abundant. Consequently, there was no conscious process of biomass regeneration.

On a national scale, woody biomass remains abundant throughout much of Africa, Asia and South America except for countries that are predominantly semi-

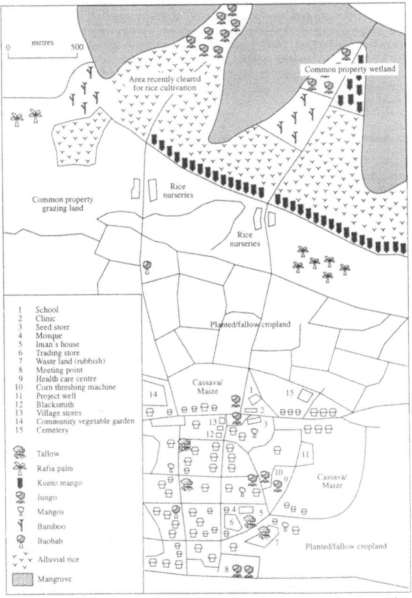

Figure 2. Sketch map of a representative section of the local wood resource base

arid. Such national abundance suggests to national policy-makers an absence of problem. This is untrue. Current management systems are based on local not national practice. Current problems are local not national. It follows that new solutions must be local.

This chapter has argued four interlinked arguments. First, woody biomass products are extensively used in local households for a wide variety of functions. Second, because people use wood, it is not surprising that they understand the importance of using and maintaining woody biomass supplies. Third, local people have evolved a system of woody biomass systems, preserving, enhancing or building new tree landscapes. Finally, local people can build these landscapes precisely because they understand the technologies of tree regeneration and tree management. In short, they have the practice which professionals must help make perfect.

Foresters need to develop skills to make quick analysis of elements in landscapes and their relation to the inhabitants.

Traditional versus farm nurseries

It is often assumed that, for the propagation of trees, nurseries are indispensable. This is questionable. Farmers may have developed other forms of tree propagation including forms where a nursery is unnecessary. (See the example of tree regeneration by farmers in the Kenyan Highlands Box 12.)

The identification of local scientific knowledge and information is a difficult process. For example, foresters are trained to start nurseries during the dry season so that the seedlings are planted at the beginning of the rains. It is a logical and well-developed system, but requires water and nursery investment. In contrast, farmers start to raise seedlings at the beginning of the rains, allowing every farmer to have his or her own farm nursery without major investment.

In principle three types of nurseries are possible:

- Central nurseries
- Community nurseries, and
- Private nurseries on farms.

Every option has its own advantages and disadvantages. These are summarized in Box 13.

In principle, nurseries with a participative character (community nurseries and private farm nurseries) are the best option, because the farmers are involved with the whole process of tree propagation and can introduce their local knowledge on tree propagation. Community nurseries are only useful when the villagers have a strong level of motivation and organization. It is not possible to succeed with forced co-operation or participation.

For example, the Peru community forestry project during the first years only focused on approaching the whole community. It was not allowed to work with groups of interested farmers or individual peasants. The idea was that the existing communal structure and organization had to be strengthened by working in a communal way. In some communities this approach worked well, in particular the communities in remote places where the communal structure was still strong. In other places, for example in communities near important roads (where modern market economy had already penetrated), or in communities with a weak internal organization, the communal approach failed. In later phases of the project it became

Box 12 Tree regeneration by farmers in the Kenyan Highlands

Under favourable climatic conditions, farmers in the highlands of Kenya have developed a range of different methods to regenerate trees and shrubs these include:

1. Planting of seedlings raised in nurseries
Farmers obtain seedlings from official government nurseries but often raise seedlings themselves in an on-farm tree nursery. The most important species raised on the farms are eucalyptus and cypress. Sometimes grevillea, fruit trees or local shade trees are grown.

Farmers start their nurseries at the onset of long rains. Doing this makes the watering of the nursery unnecessary. The nursery is established in a protected, shady place such as under banana plants or in a vegetable garden.

The most common nursery type is small (for example, 60x100cm) raised seedbed on which branches with mature pods are laid on top. The seeds drop on top of the seedbed if the pods dry. The seed germinates easily in the moist atmosphere under the branches. Later, the branches are removed. When the seedlings are big enough they are either transplanted directly (usually naked rooted), or wrapped in banana fibres that keep the root moist. The latter is usually done if the seedlings have to be carried elsewhere. Sometimes farmers use containers to raise the seedlings, in particular the more valuable species like fruit trees polythene tubes are used, but also local containers such as old tins, school milk packets, old wash basins or old pots.

2. Direct sowing
Several trees and shrubs are easily regenerated by sowing seed, either in individual holes (fruit trees), in lines (hedges or terraces), or broadcast. Commonly sown species are black wattle (*Acacia mearnsii*), *Sesbania sesban*, Mauritius thorn (*Caesalpinia decapetata*) and *Cassia siamea*. Direct sowing is not suitable for all species. There must be sufficient seed and the species must be hardy enough to survive the vulnerable seedling stage with little care and protection.

3. Use of wildlings
Many trees and shrubs regenerate spontaneously from seed or root suckers. Farmers often make use of these, so-called wildlings, by protecting them where they grow, or by transplanting at an early stage. Compared to direct sowing or raising seedlings in nurseries, this is a simple and fast method. Common species regenerated in this way are: *Markhamia lutea, Sesbania sesban, Grevillea robusta* and *Croton spp.*

4. Coppicing
Coppicing is regrowth from stumps of trees or shrubs that have been cut. It is a popular method because it saves planting of new trees after wood harvest. The ability to coppice, the vigour of regrowth and the frequency a tree can be coppiced, are species-specific, but also depend on the height of cutting, age of tree, the number of shoots that are allowed to grow through and the management of the trees (site, spacing, tools).

Box 13 Advantages and disadvantages of differents nurseries

Traditional nursery

Government nurseries run by the Forest Department, or other tree nurseries run by NGOs, are the standard way of raising seedlings in most programmes. They tend to be large (100,000 seedlings or more a year) and well-equipped, with pumped water, fencing, stores, an office, germination sheds, tools and equipment and so on. Seedlings are generally provided at highly subsidized prices, or free.

Advantages:
• Good quality control possible
• Easy to supervise
• Justifies investment in pumped water on dry sites, and
• Can act as centre for extension and research.

Disadvantages:
• Very expensive (labour, equipment, materials, transport), and
• Seedling distribution limits area that can be served.

Farm nursery

Tree nurseries on farms are a traditional feature in some areas. They tend to be very small, producing perhaps 100 seedlings a year, mainly for planting on the farm. Sometimes private farmers grow tree seedlings for sale in larger nurseries, producing thousands of seedlings every year.

Advantages:
• Extremely low cost alternative
• Distribution problem solved
• Maximum control in hands of farmer, and
• Little need for outside help.

Disadvantages:
• Seedling quality depends on farmers' skills, and these may be poor
• Water supply a common problem in dry regions, and
• Nursery management may take place during the peak period of agricultural season, and therefore is likely to receive little attention.

Community nursery

In some areas farmers may have a strong tradition of organization. Projects can choose a communal organization to have one village or community nursery established instead of several small farms.

Advantages:
• See advantages of the farm nursery, and
• Support for nursery is easier to organize than with farm nurseries.

Disadvantages:
• See disadvantages of farm nursery
• Good organization is a must, and
• Distribution of the product among the farmers may be problematic.

normal to work with groups of interested farmers instead of the whole community. Often this solution was proposed by the farmers themselves. This raises the question who has the most common sense, the forester or the farmer?

Seedlings or seed? — simplifying tree establishment

It is difficult to change direction from forestry nurseries to small and irregular farmers' nurseries. Individual farmers' nurseries will generally have a much wider impact than the expensive and centralized foresters' alternatives. Furthermore, the farmers' nurseries were more sustainable because they do not require constant inputs from outside.

Foresters must learn participation. This means they must start with local people. Such a starting point requires that local people have access to planting stock. There is a technical debate about planting stock — seed or seedling — but the outcome of the debate is not technical. The outcome of the debate is managerial, it determines the participatory process. There is never an absolute answer in rural forestry but there is best practice. Increasing such best practice would emphasize seed distribution not seedling distribution. Box 14 contains details of this debate.

In a farmers' nursery system, forestry projects can concentrate on the distribution of seed of new species rather than seedlings. The role of the government will, therefore, shift from a tree seedling distributor to a tree product developer. Not only will the role change, the impact will be greater because of increased farmer participation. To produce this situation, two major bottlenecks must be overcome:

- A farmer's nursery system based on available farm inputs has to be developed in co-operation with the farmers, and
- A consciousness-raising exercise has to reach all farmers informing them about the potential of trees.

The second bottleneck will be easy to overcome in case of a real shortage of biomass. In that case, the farmers will already know that there is a problem and the consciousness-raising exercise can concentrate on systematizing the already existing awareness.

In many countries the shift from central nurseries to farmers' nurseries and, the shift from plantation forestry to local forestry development is a gradual one (see Box 15: various forestry strategies). It should be noted that developments do not necessarily pass through all these phases, that is the type of intervention is essentially very site-specific.

One stark conclusion emerges from all the discussion of technical issues. Technical choice implies management choice. The way a project is run is, to an extent, determined by the decision over technology. Participatory projects require a widening of technical opportunity and a realization that choosing the technical option is less important than building the participatory framework. It is a hard lesson for foresters. The forester's technical expertise is an input, but a secondary input, into the process of rural tree growing.

Technical issues: the question of species

If, as a forester, you are still reading this small volume, you will wonder when technical issues will be addressed. Do not worry, they are addressed in this chapter.

But note the position, technical issues are raised towards the end, as in the new way of conducting projects, not at the start.

It can be tempting to select tree species that are commonly planted in the area by foresters and farmers. People know these species already. The benefits that can be expected, the methods to grow them and where to plant them are all known. Seed or vegetative planting material is usually available.

Box 14 Options for tree establishment

Seed versus seedling distribution
Rural forestry projects have usually relied almost entirely on providing planting stock in the form of tree seedlings. In recent years there has been an increasing shift towards the idea of providing seeds to farmers. Advantages and disadvantages of seedlings and seed distribution are summarized below.

Seedling distribution

Seed distribution

Advantages

- Can be raised by training staff under controlled conditions
- Special treatment can be provided for difficult species
- Seedlings can be raised in containers: can be kept by farmers until proper planting time
- Better able to stand competition from weeds (compared to direct sowing), and
- Quicker reaching a size where they are safe from drought, fire and browsing (compared to direct sowing).

Advantages

- Cheaper to produce and distribute
- Easier to store and transport
- Timing of distribution is less crucial
- Greater transfer of responsibility to farmers
- Can reach many more farmers with the same staff and means, and
- Greater flexibility for farmers: they can use the seed when it is most convenient.

Disadvantages

- Expensive to raise, especially in government run nurseries
- Bulky and expensive to transport
- Easy to damage during transport
- Timing of distribution is crucial, especially if rainy season is short, and
- Good nursery management is essential, otherwise quality of seedlings are low.

Disadvantages

- Unsuitable in cases where seed is expensive or hard to get
- Unsuitable for some species that require special seed treatment
- Lower survival rate with direct seeding, especially under dry or difficult site conditions, and
- Local people need experience or training in nursery techniques if they are to raise seedlings by themselves.

Box 15 Various forestry strategies

Phase 1: Plantation forestry

Characteristics
- Aimed at timber production or watershed management
- Serves regional or national interests in the first place
- Local population is seen as a threat to the forest
- Small number of tree species
- Monocultures, and
- High level of technology and inputs.

Trends, problems, limitations
- Increasing pressure on the forests from surrounding areas for tree products and agricultural land
- Increasing problems with the conservation and management of natural forest and plantations
- Worldwide growing awareness about deforestation and rural energy crises, and
- Growing political pressure on forestry institutions to change their focus.

Phase 2: A careful look outside

Characteristics
- Forestry institutions try to provide services to rural population in the form of tree seedlings
- For this purpose seedling production is decentralized but remains within forestry infrastructure, and
- Species choice is still limited to forest species.

Trends, problems, limitations
- Growing awareness among the rural population
- Logistical constraints concerning seedling distribution: only limited areas can be served, and
- Growing interest among non-forestry institutions to start working on rural tree planting

Phase 3: Moving out of the forest

Characteristics
- Stronger orientation towards rural areas
- Tree nurseries are set up in the rural areas by both forestry and non-forestry organizations
- Growing variety of tree species, and
- Nursery techniques in principle the same as in forest nurseries.

Trends, problems, limitations
- Nurseries have a relatively small impact area
- Growing demand for multi-purpose trees, and
- Growing demand for tree management systems aimed at specific products.

Many rural afforestation programmes are in this phase. In spite of all efforts by government and other organizations, the impact of all these nursery activities is relatively small.

Phase 4: Looking at farm forestry

Characteristics
- Own initiative of rural population in tree-planting and tree management becomes recognized and is investigated
- Growing awareness that woody biomass is more than trees alone: hedges, bushes, crop residues and so on
- Use of local materials, not dependent on external inputs, and
- Flexible with respect to species choice, planning and production level.

Trends, problems, limitations
- Species choice limited to locally known and available species
- Lack of knowledge about new species and management systems
- Lack of knowledge about un(der)-utilized local species
- Lack of knowledge about farmers' priorities
- Lack of trained staff, and
- Logistical problems.

Phase 5: Local forestry development on the basis of existing local farmers' infrastructure

Characteristics
- Shifting attention from seedling production in central nurseries to forestry extension activities
- More intensive contact with farmers in developing better tree and woody biomass systems
- Better assessment of local needs and priorities
- More emphasis on local regeneration methods and management aspects
- More focus on production of specific products, and
- Establishment of linkages with other rural development sectors.

Trends, problems, limitations
- Collaboration and competition between different rural development institutions working on agroforestry
- Outreach of agroforestry knowledge from forestry institution to other rural development institutions, and
- Resistance from within organization to hand over certain tasks.

However, in spite of these advantages, will these species be able to fulfil a particular role in the project? For example, are trees that produce poles good fodder producers, or are fast-growing timber trees the right trees to plant on terraces for erosion control? Several factors must be considered during the process of species selection including:

- The role trees have to play is determined by the products that are required, for example building material, fuel, fodder, fruits, or certain services or environmental functions, such as improving soil fertility, controlling soil erosion, providing shade, ornamental functions, fencing and boundary demarcation
- The environmental conditions which limit low-cost tree planting interventions. What is rainfall, temperature, soil and hydrological conditions, the occurrence of pests and diseases, the competition with existing vegetation and the pressure of wild life?
- The land use systems that define the primary role of trees. Are trees in grazing systems, in intercropping systems, or in farm woodlots and home gardens?
- The socio-economic and cultural factors that determine success. What exists in terms of belief taboos and beliefs regarding specific species or trees in general, ownership and control over trees or specific species, the market value of certain trees, and traditions regarding tree planting and tree management?
- The possibilities for distribution of project output. What is the availability of seed or vegetative planting material and what is the availability of simple propagation methods?
- Variation within single species. Is there considerable variation in growth characteristics within one species, between provenances from different regions or between sub-species or varieties?, and
- Indigenous versus exotic species. What indigenous species are often overlooked because there is little information about them?

Quite simply, tree selection is difficult. Here is an example. In the Kenyan Woodfuel Development Programme, the objective of the project was to improve the supply of firewood for household consumption by improved tree planting and tree management methods. In the process of tree species selection, the following factors were considered.

First, as the production of trees was for firewood or charcoal, or both, the chosen species must grow fast, produce firewood on a regular basis (short rotation), produce good-quality firewood (burn regularly, no sparks or smoke) and must be easy to harvest and handle.

Second, the species must be suited to the tropical highland climate in the area (altitude 1200–1600m; rainfall 1400–2000mm; short dry season) and the soil and hydrological conditions which were sandy and loamy, sometimes shallow and stony, but with medium fertility in a well-drained soil.

Third, the land use system is a mixed farming system with foodcrops (maize, beans, millet, sorghum) cash crops (coffee, bananas, sugar), animal husbandry and commercial woodlots (eucalyptus). Nearly all land is privatized and there is a high intensity of land use. The chosen species must fit in this system. Since there is little land left for pure plantations, cropland, farm boundaries, terraces and hedges must be used for tree planting. This requires species that are compatible with agriculture, easy to control, do not produce a lot of shade and preferably have soil improving qualities such as nitrogen-fixing and foliage that easily decomposes.

Fourth, there are the socio-economic and cultural factors. All resources are owned by men, including trees. Women have no free access to commercial trees, so only small trees and bushes can be used freely for firewood. The firewood issue has a low priority in the community — it is a hidden problem, a women's problem. But there is a strong taboo against tree planting by women. The chosen species must therefore be acceptable to men, accessible to women and have multiple purposes.

Fifth, there are possibilities for distribution. KWDP's approach has been to encourage farmers to propagate trees themselves in accordance with what farmers are doing already with eucalyptus, black wattle, and cypress in small on-farm nurseries, by direct sowing, by planting cuttings, or by collecting wildings. The chosen firewood trees, therefore, must be produced locally within a short time to facilitate quick distribution over the whole district.

Finally, what of the indigenous versus exotic species debate? The KWDP promoted both indigenous and exotic species. In general, exotic species are more easily adopted mainly because they grew faster than local species.

But why do projects favour indigenous or exotic trees? Many forestry programmes almost exclusively plant exotic species which were introduced during colonial period. These species are chosen because they are well-known to foresters. Although there are exceptions, most farmers grow only a handful of these species. It is often argued that indigenous species should be investigated and made available for tree planting to replace some of the exotics that create ecological problems. However, both have advantages and disadvantages.

Although natural forests are being cut rapidly, the number of indigenous trees and shrubs in many areas is still overwhelmingly large. The advantages of indigenous species are that they are adapted and proven under local conditions, familiar in their growth and products to local people, help to preserve the diversity of flora and fauna, unlikely to become a weed and produce locally available seeds. Their major disadvantage is their generally lower growth rate. Well-known local species are often not planted by farmers for this reason. However, it is possible that local species have passed through ages of negative selection in which rapid growing single trees of the appreciated stem-form were first removed, ensuring that the genetic resource deteriorated.

Carefully selected exotic species have a great deal to offer. Rapid growth, uniform products, and the availability of different varieties or provenances for different ecological circumstances are all attractive features.

The introduction of new exotic species can help diversify the species package and to make a shift away from species like eucalyptus. A number of promising leguminous species are being tested in many countries, for example *Leucaena leucocephala, Calliandra calothyrsus, Gliricidia sepium, Prosopis spp.* Many of these species are a different type of tree than the straight, tall forest trees that are normally advocated by foresters. They are smaller, often branchy, and have properties that make them easier to integrate in farming systems. They can be pruned or coppiced regularly, improve the soil, their leaves and pods are useful as green manure of fodder, and they can be planted, scattered or in hedgerows. Especially in situations where there is pressure on land, this flexibility and multiple use is valuable.

Eucalyptus — to plant or not to plant

No one can address the indigenous versus exotic debate without asking questions about eucalyptus. A number of the 700 different eucalyptus species, native to Australia, have been planted in many parts of the world. Whether this has been a sound idea has been the subject of considerable controversy. The debate usually focuses on the alleged negative ecological effects of eucalyptus as opposed to positive characteristics such as rapid growth or drought resistance.

The major issues surrounding the debate can be summarized with five statements:

- 'Eucalyptus lowers the ground water table' — rapid growth is always associated with high consumption of water. The question is which is most important in the specific local circumstances — wood or water?
- 'Eucalyptus causes rainfall to decline' — there is no conclusive evidence that eucalyptus has an effect on the local climate
- 'Eucalyptus increases water run-off' — water run-off under eucalyptus is greater than from grassland or low shrub vegetation. Grass cover tends to be sparse under eucalyptus, especially in dry areas, and where trees are closely spaced. This is also true for other species, but less so
- 'Eucalyptus exhausts the soil' — if planted on previously treeless sites, eucalyptus can be expected to improve soil fertility by increasing humus levels. However the effect is less pronounced than with broad-leaved species, especially leguminous trees, as eucalyptus species decompose badly, and
- 'Eucalyptus creates ecological deserts' — eucalyptus plantations do not support a wide variety of animal or bird life. Conservation of patches and corridors of natural vegetation can reduce the negative effect.

It is increasingly accepted that many of the negative effects attributed to eucalyptus should not be blamed on the tree itself but on the management system that is used. If trees are planted on well-chosen sites, in an integrated land use system where eucalyptus does not dominate, the negative effects will probably be smaller than the positive. In many parts of east Africa, farmers already grow eucalyptus for poles, timber and firewood. They plant the trees in swampy areas (sometimes deliberately to drain swamps), on stony hilltops or on the roadside and avoid planting them near agricultural fields. From their own experience, farmers know the ecological disadvantages of eucalyptus well and are, therefore, careful where to plant them.

Miracle trees?

The introduction of certain so-called miracle tree species to farmers has been very successful. An example is *Leucaena*, which is planted extensively by farmers in many countries. Miracle trees often belong to the *Leguminosae*, a taxonomic family which has many members that possess nitrogen-fixing root nodules.

The reason for the success of these miracle trees can be attributed to the following properties:

- Quick growth
- Multiple use
- Positive influence on soil quality and agricultural production
- Easy propagation, and
- Coppicing possibility.

The New Forester

An important task of the new forester is to find tree species that live up — as much as possible — to these requirements in the specific local circumstances. In specific local circumstances, some of the above-mentioned properties will be relatively more important and some less. Maybe other properties have to be added. The social forester should find out, by asking the farmers, what types of trees are needed, and what native species may fit into the ideal profile. In some circumstances, for example in high mountains in a dry climate with periods of nightfrost, it may be very difficult to find species that match to the ideal profile of a miracle tree. Improvement of the production of native species or a programme to adapt exotic miracle trees to the specific local circumstances may be important lines to pursue.

It has to be noted that the proposed procedure to be followed is contrary to the current practice in many community forestry projects. Normally the forester has a certain species that he or she wants the farmers to plant and to use. The social forester, however, asks the farmer what may be useful to the farmers and tries to find, or to develop, a species, a variety, or a production system that fits to the farmers' needs.

Much emphasis is laid on the multi-purpose character of trees. However, knowledge about the correct production system to obtain a desired end-product is often lacking. The emphasis is still on national supply, not on local demand. For example, *Calliandra calothyrsus* is known as a valuable species for fuelwood, fodder, or fencing. But little is known about how the trees should be locally managed to obtain those products in the most efficient way. This emphasis on local management requires a shift in the role of forestry in rural development.

There is a danger in relying on one miracle tree particularly the species *Leucaena leucocephala (Ipil-ipil)*. This species is susceptible to the psyllid disease. This is one of the reasons to pursue species diversification. For *Leucaena leucocephala* and three other very promising species (*Sesbania, Gliricidia*, and *Calliandra*), issues will be raised in the following paragraphs. But, it has to be observed that, while much has been published on *Leucaena*, publications on the other species are relatively scarce. Many other species might be promising, but investigations about their possibilities are not yet started.

a) Leucaena leucocephala

Leucaena is a leguminous tree which originated in Central America but which has been widely introduced in South-east Asia. Under optimal conditions it grows very fast, improves the fertility of the soil, is able to control erosion if planted in contour lines and produces a whole range of useful products, including fodder, edible pods, and high-quality wood for various purposes.

The impressive results have made the species extremely popular in many parts of the world. Expectations have been very high. But in many cases it has failed to perform well. Two issues play a role. First, the species has a number of varieties which can either grow a single straight stem or be bushy. Some varieties are known for their relatively large leaf production. Choosing variety is important. Variety must be matched to project purpose. Second, *leucaena* is often disappointing if it is planted without regard of local site conditions. It is often assumed that the species can grow well anywhere. In practice, however, it will not grow well under the following conditions:

- High altitude; it prefers altitudes below 1000m

- Drought — it requires at least 800mm/year rainfall to grow well
- Waterlogged conditions
- Acidic soils
- Under shade, and
- Heavy competition from weeds.

It is often assumed that leucaena can be intercropped with maize and other crops without lowering crop production. Sometimes it can, because of its soil improving ability through nitrogen-fixation, but only with good management. This good management requires regular pruning or cutting back to reduce competition with crops, incorporating leaves in the soil or use them as mulch, spacing the trees and proper timing of planting.

b) Sesbania spp.

Leucaena is not the only modern miracle tree. Fast growth, wide soil tolerances, multiple uses, and a wide range of growth habits and adaptability make *Sesbania spp.* a treasure of diversity and versatility.

The genus *Sesbania* contains about 50 annual and perennial species spread throughout the tropics and subtropics. Perennial woody sesbanias include *S. grandiflora* (pantropical), *S. javanica* and *S. formosa* (Australia), *S. arborea* (Hawaii), some little-known African species, and the *S.sesban* group (Africa and Asia), *S. bisponosa* (India) and *S. emerus* (Central and South America). Africa has the greatest species diversity within the genus and Asia has the greatest utilization of *Sesbanias* in cultivation.

Sesbanias belong to the family of *Leguminosae* (legumes). They are small trees or shrubs. Some can grow up to 6m high. *Sesbanias* form root nodules that are capable of fixing atmospheric nitrogen and are adapted to diverse, difficult sites. Many species tolerate soil alkalinity and salinity and can be used in soil reclamation programmes. Unlike other legumes, sesbanias survive waterlogging and flooding and continue to fix nitrogen which makes them useful in lowland rice cultivation areas and seasonally flooded lands. There are indications that low levels of phosphorous do not severely inhibit growth of certain species. Different *Sesbania* species are adapted to different climatic conditions. For example, *S.grandiflora* is a tree of the humid and semi-humid tropics up to about 800m above sea level. *S.sesban* can grow in both semi-arid (at least to 350m above sea level) and humid climates up to about 1200m above sea level.

Sesbanias have diverse uses in cropping systems. As green manures, they are unsurpassed in moist environments and grow very rapidly in hot climates. In many areas, *Sesbania* species are already cultivated by farmers, either by direct sowing or by tolerating and protecting wild seedlings. They are often grown within or alongside cropland, either intercropped or as fallow vegetation. Farmers claim *Sesbanias* do not effect crop yields and maintain, or even improve, soil fertility.

Agroforestry applications of perennial *Sesbanias* have only begun to be explored. Perennial *Sesbanias* are planted in Indonesia and India as shade or windbreaks for vegetable gardens and to support black pepper, cucurbits and betel vines.

Flowers and young leaves of *S. grandiflora* and *S. sesban* are eaten as vegetables in Asia. Leaves and green pods of both species, and other sesbanias, can be lopped as fodder for ruminants. *S. sesban* provide better yields but slightly lower nitrogen levels (between 25 and 30 per cent) than *S. grandiflora*. Some *S. sesban* varieties are able to produce more than 20t/ha/yr in dry matter. If lopped, the shrubs sprout vigorously and can be lopped again after a short time. Both perennial and annual species can provide for fuelwood and pulp fibre. *S. sesban* fuelwood plantations in Kenya yield up to 30 t/ha/yr of air dried firewood. In India, an annual production of about 75 t/ha/yr has been recorded.

c) Calliandra calothyrsus

To learn more about this species, some seeds of *Calliandra* were given to each of the 1700 forest guards scattered in rural Java. The resulting trial plots demonstrated that *Calliandra* would be most suitable for village fuelwood production in humid areas at medium elevations (250-800m above sea level). Below 400m, *leucaena* was the best species and, from 800 to 2000m, *Albizzia montana*, an indigenous tree legume, was most successful. In many areas, *Calliandra* is the source of products that used to be extracted from the disappearing natural forests. Plantations can be established by direct seeding. The tree is fast growing: within 12 months the plants may be 3-5m tall and 5cm in diameter at stump height. It may be used for fuelwood, for soil improvement, fodder, honey and for ornamental uses. In hedges, in Costa Rica, it has produced 3.4kg/year fodder for each metre of planted hedge.

d) Gliricidia sepium

Gliricidia sepium originates from Central America. The species may be established by direct seeding, seedling transplants or by planting of vegetative cuttings. *Gliricidia* may be used for living fences, alley-farming, land rehabilitation, medicine, rodenticide, pesticide, fodder production and for fuelwood. In Samoa, it yielded 30 t/ha/yr of fresh fodder, intercropped with taro.

Agroforestry production systems

The most important factor to consider in attempting to promote a new production system is that a production and management system already exists on the farm. Coming to terms with this, and attempting to build on or improve the existing production system, is critical to project success. Again, it is a question of listening to the local people and looking at the structure of the production system. There are already short-, medium- and long-term rotation systems in existence with which the farmer is familiar.

A key issue in production systems analysis that is again a consequence of inappropriate training, is the assumption that the value of trees is proportional to age. Foresters, from the classical school of forestry, have a perception that tree value increases with age and size. This tends to be the case in industrialized countries where tree size increases tree value, manageability, and economic return. This tendency is not absolute. In the classical view, trees are not cut at their maximum volume. They are cut when the invested sum in plantation, land and management actions reaches its optimal internal rate of return. In general, this stage is reached when the trees are relatively old and have considerable volume. In developing countries, however, farmers are likely to put a higher value to small trees and shrubs. Their time frame and economic rationale for tree growing is therefore quite different.

Figure 3 compares the two different valuations for developed and developing countries. In the industrialized countries costs are minimized by costs of harvesting and processing being balanced against maximum tree size, age and weight. In developing countries costs are minimized when trees are harvested young and labour costs (in terms of time) are low. The different systems also relate to end uses of tree products. In developed countries, end uses may be paper, pulp, fibre and building material, contrasted to fuel, fodder and poles in developing countries. With this in mind, the social forester can design projects to facilitate individuals' actions that maximize the benefits to the farmer. First the forester must ask what the farmer wants to receive as a benefit?

The following questions need to be considered:

- What is the demand for the specific tree product?
- What is the rotation system that can most rapidly deliver the products and how well will they fit into the land use system?
- What tree species are best suited to these land use systems?
- What are the available resources? Where are the bottlenecks: is it, for example, land or available labour?
- Is it possible and useful to improve the raw material the tree produces, so that a product emerges with a higher value?
- Is it necessary to develop a marketing strategy for the product?
- Is the new system simple so that rapid and spontaneous dissemination of the technique is possible?, and
- Are the proposed changes gradual?

The word tree has so many different connotations that it is a real obstacle to the development of wood production. It has to be recognized that the tree is just as viable to produce as is a grass or crop. But a tree is not directly related to the time period required to mature — some species take more than 100 years, others need only 6 to 12 months to be harvested as a wood product, although trees are often assumed to be old and slow growing. For farmers with smallholdings, time spans of eight years are generally beyond their livelihood limits. Most forestry knowledge, however, concentrates on wood products with production periods of more than eight

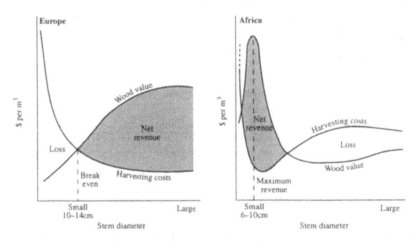

Figure 3. Wood value as a function of stem diameter — developed and developing wood production systems.

years. External agents have almost always offered or implemented long-term rotation systems aimed at producing large trees. Indeed, subsidies are offered for these projects. There has been, and still is, a gross misunderstanding of time in the tree growing process.

There are two basic departure points for building successful social forestry projects. They are:

- The use of a tree is more important than the method of establishment, and
- There are always trees on farms.

Once these two points are understood, an external agent can begin developing the production system, by building on the trees that are already there. An improved short-term rotation system can begin by, for example, extending the hedge. Such interventions, close to houses, allow farmers to observe, experiment and research new and old species.

A scheme to categorize agroforestry production systems

The following section contains details of agroforestry production systems by rotation systems (see Box 16). The schemes are based on experiences obtained in Kenya. Such a scheme may serve as an example, but for local circumstances the strategy must be shaped from local experiences and not imposed on local people.

The division into short, medium, and long rotation systems may be useful in order to systemize the available local information. The actual system (for example hedgerow, windbreak), species and length of rotation will depend on local circumstances, needs and possibilities.

Short rotation cycles.

People are essentially short-term creatures, their vision of the future is often clouded by the circumstances of today. This is why quick and beneficial results are very popular. The short-term rotation system provides fast results. It also gives valuable results.

Short rotation cycles typically last for one to two years in production, after which they are harvested, thinned or coppiced. They provide a multitude of benefits (including fuelwood, fodder, mulch, shade, boundaries and small building matter) and are essentially, in terms of the new paradigm, the most successful, beneficial and valuable tree management systems in the majority of the developing world.

This may surprise and indeed challenge the position of most foresters and perhaps they will contend the issue. However, the message within this chapter is not to lay down another blanket approach but to offer alternatives, backed up with evidence to suggest methods of overcoming the problems of contemporary forestry. Each forestry situation is different. No specifics can be laid down for intervention. What must be realized is that the knowledge base is often there already to be built upon. The rationale is:

There is not a **right** answer;
there is simply **an** answer.

Short rotation cycles are the methods that usually maximize benefits. The labour costs associated with harvesting, collection and transport are low and the benefits of the products gained is high.

Box 16 Agroforestry production systems

System	Tree species	Products
1. **Short rotation systems**: Benefits can be obtained in one to two years		
Hedgerow systems:	*Sesbania sesban*	Firewood
• Alley cropping	*Leucaena leucocephala*	Fodder
• Boundary hedges	*Calliandra calothyrsus*	Twigs for basket
	Gliricidia sepium	granaries
• Soils cons. hedges	*Sesbania grandiflora*	Small building matter
(contours, terraces)	*Sesbania bispinosa*	Mulch and green manure
• Fallow vegetation	*Sesbania spp.*	Mulch and green manure
	Acioa barteri	Firewood
Trees in food and cash	*Erythrina spp.*	Shade for crop
crops, e.g. coffee, cocoa,	*Cassia spp.*	Firewood
maize, beans	*Leucaena leucoc.*	Fodder
	Gliricidia sepium	Mulch and green manure
Woodlots, wind-breaks	*Sesbania sesban*	
and so on	*Calliandra cal.*	Firewood
	Mimosa scabrella	Wind protection
	Cassia spp.	Nectar for bees and so on
2. **Medium rotation systems**: three to six years		
Woodlots, wind-breaks	*Eucalyptus spp.*	Building poles
trees on boundaries	*Acacia mearnsii*	Fencing material
	Cassia siamea	Firewood
	Mimosa scabrella	Wind protection
	Melia azedarach	Soil conservation
	Prospis spp.	Fodder
Trees combined with	*Inga spp.*	Poles
food or cash crop	*Cassia spp.*	Fencing material
	Albizia falcataria	Firewood
		Soil improvement
3. **Long rotation system**: six years and more		
Multispecies mixtures:	*Artocarpus spp.*	Timber
(home-gardens, tree-	*Albizia falcataria*	Fruits
gardens)	*Mangifera indica*	Spices
	Cocos nucifera	Poles
	Parkia spp.	Fencing material
	Swientenia spp.	
Woodlots, wind-breaks,	*Eucalyptus spp.*	Timber
shelterbelts, trees on	*Cassia spp.*	Poles
boundaries etc.	*Grevillea robusta*	Firewood
	Cupressus lusitancia	Charcoal
		Wind erosion control
Trees in cropland or	*Albiziz falcataria*	Poles
scattered on the farm	*Albizia spp.*	Timber
	Grevillea robusta	Charcoal
	Casuarina equisetifolia	
	Alnus acuminata	
	Cordia alliodora	
	Markhamia lutea	
	Croton spp.	

Consider fuelwood production where the labour costs associated with large trees are high. What is the point of growing large trees when the product, fuelwood, has to be small, to use in a stove. It is far better to grow small, harvest small and burn small. There is the added factor that heat for cooking is needed relatively quickly, with a large heat output, and must be easy to light. Wood that is small, quick to light and good for heat is young wood. Old, cut wood burns for a long time, is in large pieces and is cumbersome to work with. Small is not just beautiful — it is useful.

In medium rotation systems, production is aimed at products that cannot be produced in shorter rotations and for which a long rotation is not necessary. They are used for producing harder materials for fencing, poles and roofing, which need some years to grow, and for producing shrubs for wind protection. Certain species may not grow well when cut on a short rotation basis. Therefore, these specimens must be managed in medium rotation systems. For example, in the Peruvian Andes management of the native species, like *Buddleia spp*, still exists. In some villages, every three or four years, the trees are cut at 1m height. As a result of this, very erect sprouts emerge, which can be used for roofing after three or four years (local name: *chacclas*). The quality of this material and the value of production is better than *Eucalyptus globulus* could produce.

In long rotation systems, the planted trees are cut after a relatively long time. The farmer will choose a long rotation because he or she needs a product that cannot be obtained from short or medium rotation systems (for example, beams for house construction, large trees for wind-breaks; shelterbelts;) or he or she uses the trees as a means of saving money. In the latter case, the tree is cut and the timber sold at the moment the farmer needs money. In this case, the farmer is not interested in an optimum economic return because, if this was the case, a shorter rotation would have been chosen. A special case is fruit trees. From a forester's point of view, these trees may be grown in a long rotation cycle, but for the farmer the most important point is that every year the fruits can be harvested.

For example, in Columbia, in the moist tropical lowlands of the Chocó, the local farmers have developed a system to manage natural regeneration of cedro (*Cedrela odorata*). Seedlings of this species emerge during a fallow period on agricultural lands near the river Atrato. Selected thinning of the wildlings is executed when the land is used again for the plantation of plátano (*Musa spp*). In the first phases, the young trees are surrounded by high weeds, a situation which protects them from being attacked too much by the *Hypsipyla grandella* top borer. After some 20 years, the trees can be harvested and are often sold to merchants in order to get an additional cash income. Another example, in Peru, is where the State forest service introduced *Eucalyptus globulus* to the Peruvian highlands. Farmers have planted this species around pasture near their houses and in woodlots. The species is used for firewood. It is especially liked because of its erect bole. Native species do not produce this erect bole. The trees are allowed to grow high and then are used for the construction of houses.

This chapter has stressed that woody biomass management is fully integrated into people's lives. There is a purpose behind managed landscapes, created by the indigenous knowledge base and perceived need for biomass products. Perhaps, next to this recognition, the most important aspect to understand is that farmers have definite time-scales for the development of woody biomass. These time frames relate directly to the need of different wood products and the farmers' understanding of the best way to maximize benefits with minimum costs.

Understanding these factors means that, for the social forester, most of the information for improving the woody biomass system is already present. The social forester's main task is to help design local management systems in time-scales that fit in with the local need for biomass products.

5. The Practicalities of Intervention

The new forester in action

Arising from the conflicting aims of the classical and social forester come two critical points. First, there is a need to adjust the classical forester's attitude so that he or she can begin to tackle the problems of social forestry. Second, this adjustment must recognize the range and value of local people's experience where they protect, enhance or build new woody biomass landscapes as an integrated part of their livelihood systems. By itself, this recognition is not enough. What is needed is a participatory development approach that will build a sustainable biomass programme by involving the forester with local people.

Already, there have been hints about how to tackle the problem. First, there is a need to turn the administrative process inside out. There is a need to reverse the normal delivery channels for forestry inputs, to make the priorities of intervention, and thus research, be community, not professionally, led. To build such a participatory process requires:

- The development of rural forestry technology that is drawn from existing experiences in particular ecological zones, and
- The development of new routings for those technologies so that they diffuse spontaneously since spontaneous diffusion is the measure of successful development strategies.

Spontaneous diffusion here means the transfer of knowledge and understanding between people without passing through formal structures. Formal structures include classrooms and lectures by extension workers. More important structures for participatory development are the conversations at the village well, over a shared glass of home-brew and walking to the fields — the non-controllable structures. This participative process requires both an interactive generation of small forestry technologies adapted to local environments and the development of organizations that can allow local people experimental capacity.

Such an approach, of course, has its dangers. Results are not generally applicable across wide areas but the methodology is. Local knowledge can be misplaced leading to difficulties when extrapolating local practice into development programmes. Most important, social structures and organizations might, within themselves, preclude a participatory approach. But if rural forestry is the goal, there is nowhere to start except with the local people. With them, new methodologies of intervention can be built.

However, there are certain precautionary measures that can be taken to judge the possibility of a participatory programme. First, the external agent must accept that he or she is locally ignorant and that expertise limits rather than expands common sense. Second, there is no need to rush into rural forestry solutions. The problems have been there for a long time and the local people realize that solutions will take time. Third, if local social structures are not corrupt, build on them rather than setting up an invasive administrative structure. Finally, the participatory forestry process is meant to enhance the local communities' capacity to experiment and innovate so their livelihood system is sustainable. It is not meant to enhance the reputation, and the earnings, of the expert.

A model for participatory development: thinking through the process

Participatory development by definition, should be a response to a problem that is voiced by an individual, a group or a community. The process of participatory development should not be indiscriminately applied using some noble justification.

Advocates of a holistic approach to participatory development perceive it as a process of response which combines social (human development), political (eradicating inequality) and economic goals. Within this perspective, the most critical variables that determine success are the way in which the development occurs, in terms of equity, and exactly who is involved and how. This suggests that any development initiative needs to have the immediate needs of the relevant population as its starting point. Perceptions, roles, goals and interests of different actors need to be clarified at the start when organizational and social relationships are given as much thought as the technical process. The community-based institutions and organizations through which effective participation will take place need to be identified (see Box 17). Social process is crucial to fostering additional participatory development initiatives.

One of the most critical factors in designing the social strategy of agroforestry programmes is the identification of the unit of social organization able to carry out the programme and the definition of the conditions under which this unit can act

---

Box 17 Example of selecting groups

In the Dhaulapur Project, in the Himalayan Mountain region of India, the first step in promoting social organization and agro-ecological development was to make a list of all types of existing social groups and organizations in the area of operation, regardless of whether they were officially registered or not.

Discussions were organized in the villages with members of the various groups and organizations to understand the traditional systems of organization: objectives, social standing/perceptions of the organizations, social and ethnic profile, communication, formal and informal leadership, and activities.

This inventory resulted in a profile of each organization or group indicating the areas in which it could become an active partner, what participation it could mobilize and to what degree it could contribute to sustaining the process of technology development.

---

effectively. Many recent or on-going agroforestry projects have lumped together, under the broad heading of social or community forestry, different objectives with vague or unfocused appeals to various heterogeneous or undefined populations.

Operationally, it is not only a challenge but an absolute necessity to disaggregate the broad term people and to identify precisely who and how: what units of social organization can and will do afforestation, and which social groups and definable units can act as sustaining and durable social structures for long-term production and management activities.

Such units of social organization can be:

- Existing social units, such as individual household or tightly knit kinship group/subgroup
- Groups organized purposely to plant, protect and cultivate trees, and
- Groups established for purposes other than agroforestry, but which are able to undertake forestry-related activities as well.

Forming enduring units of social organization is particularly important in the case of agroforestry strategies — given the relatively long duration of a production cycle — in comparison with agricultural crops. Even small, self-managing groups enhance the individual productivity of their members. They increase the cumulated impact of the individual contributions and enable members to perform works and achieve goals that might not be attained by each acting separately.

In agroforestry, self-managing groups acting as economic agents can achieve significant economies of scale in several respects:

- Primarily (but not only) with respect to labour required for tree planting and cultivating
- In labour for harvesting and transporting, and
- In bargaining more effectively than individuals when selling the harvest or when negotiating with authorities.

Furthermore, some specific technological needs or constraints may be more easily solved by groups, particularly guarding and protecting tree plantations against theft, fire or destruction by animals.

The need to identify or establish social units capable of collective action introduces another sociological dimension in agroforestry development projects and into the work of forestry departments. If properly conceived, social agroforestry projects can become a mechanism for encouraging and forming groups, thus building up the social capacity for development. Helping users to organize themselves into groups, and to undertake production and management functions in agroforestry, will restore the balance of the participation equation: the users of forests and forest products act as the primary producers and decision-makers, and the forest department will participate in their activities, rather than the other way around.

Establishing a functional social group means much more than simply lumping individuals together in an artificial entity given the label 'group' on paper. It implies a process of selection or self-selection of the members, the willingness to build the members' perception of both self-advantage and co-responsibility, and the establishment of an enduring intra-group structure with well defined functions.

At the same time, however, the building of groups has to address certain complexities resulting from the actor being a group of farmers, rather than an individual farm household. These issues include joint dependence over a piece of land; group management, labour allocation and monitoring; and, probably the most sensitive, benefit distribution. Organizing and promoting groups as units of social organization for social forestry programmes means designing clear social arrangements for tenure, management and distribution, arrangements that are known, implemented and adhered to collectively.

Unfortunately, there is a large difference between theory and practice in participatory development. Many projects are now called participatory or social but, in fact, are not. This may partly be due to the fact that 'social' or participatory

Box 18 Trigger factors — participation in agroforestry

Participation in agroforestry begins when individuals, either singly or collectively, begin to plant trees. The impetus to plant a tree may come from various sources but essentially the individual must want to plant a tree him or herself. The trigger factors have been very important to understand because, by understanding these factors, it is possible to understand the rational reasons for tree planting.

The most common type of trigger factor has been when farmers have seen successful examples of tree growing, such as seeing a neighbour farmer harvest and sell timber for profit. In general, trigger factors are a response to a role model or a model of best practice. For example, in Nicaragua, a local farmers' co-operative approached an NGO field officer about improving agricultural production. The field officer explained a method whereby half an acre of land was given over to growing Papaya fruit which theoretically would bring a return of $2500 a year. The co-operative was not convinced of the method and wanted to grow beans and maize rather than involve itself with a foreign farming technology. The field officer employed his method himself while members of the co-operative looked on and observed. After a time, when the farmers saw his success, they came back to the field officer for further discussions. Thus, the trigger factor here was the model of best practice provided by the field officer carrying out his advice with his own resources.

Access to tree seedlings has been an important trigger factor, since many farmers have planted their first trees after being given a seedling from a friend, neighbour or nursery. Advice by an extension service has generally not been a trigger factor.

Distribution of tree seedlings from nurseries has been important, not because of the physical impact of the relatively small number of trees, but because it exposes many individuals to a tree growing experience. Anyone thinking about growing trees must be able to see success elsewhere. Many of the most successful tree planters today point to a certain tree which they had planted as children which in later life, has encouraged them to plant trees on a large scale.

People remember:

- Twenty per cent of what they hear
- Forty per cent of what they hear and see, and
- Eighty per cent of what they discover for themselves.

programmes cannot be laid down in a systematic schedule like an engineering project. Methodologies can be identified but no specific goals can be given — this is the nature of social activity. Moreover, the social forester cannot operate in a sectoral framework as classical forestry does. In social forestry, foresters need agriculturalists, planners, sociologists and anthropologists to help them to design and test a combination of technical and social approaches to build coherent development projects.

It is possible to give a flavour of how a participatory development programme could be planned. It is an interative process, taken out of the office and laboratory, into the field. Getting started requires getting to know people and their habits — social knowledge precedes implementation of anything. The design of experiments reflects the priorities of local communities and the experiments are not written in stone, as technical packages, but are local trials. These local trials usually develop from a impetus or a trigger factor based on a perceived need (see Box 18). Most importantly, the process is not a one-off attempt but requires sharing of results, including failure, with other groups and so requires a mechanism to consolidate networks for information exchange.

In Box 19, the seven-step process is summarized. In connecting on this process, examples will be drawn from the Kenya Woodfuel Development Programme to highlight the participatory development model.

In the selection of a project site, the starting point is to find an area with trees already in production. Understanding the existence of such tree coverage is the key for further woody biomass development. When there is an explanation of current successful tree management practice, areas with less tree coverage can be selected.

It was for this reason that, in Kenya, the Kenya Woodfuel Development Programme started its work in Kakemega (Box 20). From earlier studies, it was suggested that a shortage of fuelwood would occur in the area in the near future. Based on supply and demand analysis, it was argued that urgent action was needed to secure future needs. Travelling through the area raised doubts about the outcome of the earlier studies. Trees could be found on most farms. Scepticism existed among the local people that there was a wood problem. Why start in an area like Kakemega when many areas in Kenya, especially the arid and semi-arid areas, were devastated by tree removal and soil erosion?

The original reason for choosing Kakemega was that it was a densely populated region of Kenya where, as land became increasingly divided into smallholdings and common property resources disappeared, one would logically expect increasing problems of fuelwood provision. In detailed analysis what became apparent was the reverse. Wood growing was, in some places, actually increasing. The focus shifted to inventorize the on-going tree multiplication techniques. Reconnaissance surveys had indicated a variety of tree multiplication techniques on farms. In fact, each locality had its own tree planting systems. During the project formulation stage, the striking observation was confirmed that, in the most densely populated areas, the density of trees was greater. Population density was not inversely correlated with tree density. Something was going on. But, whatever was going on was difficult to formulate into a fundable project proposal — ignorance of local success is hardly an attractive proposition for donors.

To build a fundable project, a traditional project was formulated requesting support for some 300 local nurseries at community-level. The major aim of the project, building on earlier studies, was to increase the fuelwood supply on the

Box 19 Seven steps in participatory forestry development

Activity	Description	Examples of operation methods	Examples of output indicators
1. Selection of the project site	• Guidelines for selection • Formulation of physical criteria • Formulation of social and cultural criteria.	• Implementing policy guidelines • Subjective selection, and • Objective selection.	• Project site.
2. Getting started	• Creating an understanding for the project among the staff • Building relationships for co-operation with farmers Preliminary • situation analysis, and Awareness • identification.	• Rapid rural appraisal • Organization resources inventory • Community walks • Screening secondary data • Community surveys, and problems and project technique	• Inventories • Protocols for community participation • Core network, and • Enhanced agro-ecological awareness.
3. Looking for things to try	• Identifying priorities • Identifying local community, scientific knowledge and information, and • Screening options, choosing selection criteria.	• Farmer experts workshop • Techniques to tap indigenous knowledge (case histories diagramming, preference ranking,- local repertoire and indicators, critical incidents). • Study tours, and • Options screening workshop	• Agreed research agenda • Improved local capacity to diagnose a problem and identify options for improvement, and • Enhance self-respect.
4 Designing experts	• Review existing experimental practice • Planning and designing experiments • Designing evaluation protocols, and • Informing organization about the progress made.	• Improvement of natural experimentation on-the-spot • Designing workshops and prompting questions slides/ videos, case histories Testing alternative • designs Farmer-to-farmer • training • Workshops with relevant organizations also active in the area.	• Experimental designs which are manageable • Protocols for monitoring and evaluation • Improved local capacity to systematically designed experiments • Creating good understanding for the field activities.

5. Trying out	• Implementation of experiments/ observation evaluation.	• Step-wise implementation • Regular group meetings • Field days, exchange visits, and • Supporting activities.	• On-going experimental programme • Enhanced local capacity to implement, monitor and evaluate experiments systematically, and • Enlarged and stronger exchange and support linkages.
6. Sharing results	• Communications of basic ideas and principles, results and PTD process, • Training in skills, proven technologies and use of experiments.	• Field workshops • Visits to secondary sites, and • Farmer-to-farmer training and hands-on training.	• Spontaneous diffusion of ideas and technologies • Enhanced local capacity for farmer-to-farmer training and communication.
7. Sustaining the process	• Creation of favourable conditions for on-going experimentation and agricultural development.	• Organization consolidation • Development of resource materials, and • Participatory monitoring of impacts on agri-ecological sustainability.	• Increasing number of villages involved • Consolidated community network organizations for agricultural self-management • Resource materials, and • Consolidated linkages with research and development institutions.

farms. Such a traditional project, the KWDP, was the vehicle within which a participatory model for social forestry developed.

Steps for a participatory process

Step I: getting started
It is necessary to create a common understanding among the project staff. Various surveys are necessary to provide baseline information for project discussion. These include:

1. Rapid rural appraisal surveys which are essentially land use surveys highlighting the role of trees. Box 21 contains guidelines for these first analysis surveys and they are relatively simple — a first analysis can be done driving through a district. The results of this survey should be

---

Box 20 Kenyan woodfuel development programme

During the late 1970s there was a rising tide of concern about the so-called other energy crisis: the depletion of the fuelwood resources on which the majority of rural people in the developing world depend for their energy needs. In Kenya, a large study carried out for the Ministry of Energy from 1979–82 predicted severe fuelwood shortages, particularly in the densely populated Highlands of the country.

The Kenyan Woodfuel Development Programme (KWDP) was one of several woodfuel projects designed to deal with these impending fuelwood problems. It is under the supervision of the Ministry of Energy and began in 1983. Its principal objectives are to develop the tree cultivation techniques and extension methods required to promote increased fuelwood production by farmers in the Highlands and, in the longer term, to hand over the running of the programme to the various government ministries concerned.

The programme was carried out in two districts, Kakamega and Kisii, both in the Highlands. Most of the area covered is at an altitude of about 1500 metres and has a rainfall of 1600–2000mm a year. It is one of the most fertile parts of the country and is officially classed as having a high agricultural potential.

Each of the two districts has about 1.25 million people. The population density is extremely high and varies from around 200 to as many as 1000 people per km2 in some places. The rate of population growth is four per cent a year. Land holdings are generally small, with some families owning only 0.2ha. There is little forest or communal land.

---

combined with available secondary data. This can be used to indicate zonal tree management systems

2. Agroforestry survey focusing on the trees to be found on farms to determine patterns of trees in farming systems. The survey should also focus on the methods of tree management in the farming systems, and

3. Cultural survey to focus on local traditions with special reference to tree planting and harvesting.

In many ways, such an exercise is not easy. To build a relationship with the local community, local staff have to be recruited to conduct the surveys. Trust is important — honest answers are needed to make a preliminary, situational analysis. Much attention has to be given to training project staff in participatory approaches. Most project staff are used to giving orders and other inputs to farmers, not to listening.

In the Kenyan Woodfuel Development Programme, the rapid rural appraisal resulted in determining nine different zones. Each had its own characteristics, based on a number of indicators including landscape, farm size, crop pattern and the placement of different trees in the land use system. In order to conduct the agroforestry and cultural surveys, school leavers were recruited and trained. These people became the link between the local population and the project. They possessed knowledge of local organizational structures and insight into the protocols for local participation. They knew local farmers, particularly those with tree knowledge, and were able to network on local knowledge. To learn from the farmers, was a difficult

---

Box 21 Guidelines for a first analysis of the woody biomass situation

Observation while driving around the project area

1. What types of woody vegetation are present in the area?
   - Forests: planted or natural; main species
   - Bush land
   - Open woodland
   - Trees in and around farming areas: woodlots, windrows, scattered trees in cropland, trees on compound, and
   - Trees in public places: markets, roadsides, along canals.

2. What is the condition of these vegetation types?
   - Well-maintained or neglected
   - Gaps because of heavy cutting
   - Natural regeneration
   - Pruning, pollarding
   - Collection of dead wood
   - Fresh stumps
   - Coppices
   - Litter, and
   - Erosion.

3. Do you observe any transportation or trading of forest or tree products?
   - Heaps of wood on roadsides
   - People transporting wood, charcoal, fruits, tree leaves and the like, and
   - People selling wood, charcoal, fruits, bark, roots, medicines and so on in markets or elsewhere.

4. Do you observe any activity related to processing or utilization of tree products?
   - Sawing or splitting
   - Burning charcoal
   - Fencing
   - Building
   - Processing of fruits
   - Basket-making, and
   - Feeding leaves to cattle and so on.

5. Do you observe any activity related to tree regeneration and management?
   - Tree nurseries
   - Transportation
   - Selling of seedlings
   - Young tree or newly planted cuttings, and
   - Pruning, clipping, thinning, clearing, coppicing.

---

experience for both the local and international staff. To leave a farm without leaving anything tangible behind was a difficult experience.

From the agroforestry survey, it was clear that farmers had knowledge of tree growing. It was also clear that this knowledge was scattered, limited to a small group of farmers. In short, there was only a small nucleus to build upon. For example, one man raised seedlings and provided seedlings to his neighbours but that was it. To use and develop this knowledge, and hidden infrastructure, was the central challenge of the project. To reach all these centres of knowledge required a breakthrough in communicating with farmers.

Step II: looking for things to try

In discussions with farmers, priorities have to be identified. The discussion should not be completely open (for example, what is the most pressing problem on the farm or in the village?), but should be focused on woody biomass. Before going into the field, the social forester should reiterate the following assumptions:

1. The outside forester must accept that he or she is locally ignorant and that expertise limits rather than expands common sense
2. There is no need to rush into rural forestry situations — the problems have been there for a long time and the local people realize solutions will take time
3. If local structures are not corrupt, it is possible to build on them rather than administrative structures since the former, not the latter, tend to be more vibrant and less invasive
4. The participatory forestry process is meant to enhance the local communities' capacity to experiment and innovate so their livelihood system is sustainable it is not meant to enhance the reputation, and earnings, of the expert, and
5. For participatory forestry projects, it is important to recognize and use indigenous knowledge. It sounds very simple but, in practise, it is very difficult to see trees from a farmer's point of view. It is still the view of agronomists and foresters that existing trees are not an integral part of the landscape — a view which differs markedly from the farmer's landscape analysis.

It is a matter of professional integrity that the social forester should treat the farmers with real respect and obey the local customs in meeting them. If the forester is going to be of any use to the community, he or she has to ask questions. These questions are important to understand the relation between trees and people, but the questions must not offend.

Box 22 has a kind of checklist of questions to be asked. The checklist contains only five major questions. These five questions are a sufficient framework to start. As long as the social forestry assumptions are kept in mind, these questions elicit knowledge to start co-operation with the community.

Step III: designing experiments

Before experiments can be designed, an analysis of the existing situation is necessary. This includes:

- That the social, cultural or material bottlenecks are determined
- That the first focus is on the development of one product only (for example,

fuelwood, fodder). Once this is successfully introduced, other products are easier to develop

- That the best way to distribute knowledge and seed is explored, and
- That experiments should be designed for farm level.

After this analysis, and prioritization, the following steps are undertaken:

1. A number of farmers have to be identified. Together with them, the experiments are designed.
2. The level of co-operation has to be determined. Is it going to be at communal- or private-level? Will discussion meetings be with a group of

---

Box 22 Five questions to ask the farmer when you visit a farm

1. How has the farm changed since you were young?
- Size
- Crop
- Livestock
- Natural forest and trees
- Planted trees
- Utilization of different tree species, and
- Land demarcation.

2. To what sources of woody vegetation does the farmer have access?
- On his or her own farm
- On communal land
- On state or public land, and
- The market (formal or informal).

3. How is the woody vegetation managed? Why in that particular way?
- Species composition
- Planting sites
- Configurations
- Timing
- Management practices, and
- Regeneration methods.

4. What benefits does the farmer get from woody biomass? What is lacking (or should be improved)?
- Food
- Fodder
- Construction material
- Raw materials for handicraft and industry
- Environmental protection
- Medicines, and
- Ceremonial and religious benefits.

5. Who is the owner of trees and shrubs? Who is responsible for management and harvesting?

---

farmers? Will each farmer execute his or her experiments on their private farm?

3. The production of necessary inputs, like seeds, has to be arranged, and
4. The organization of a mass awareness campaign must begin. Objectives of the campaign include actions:
   • To make it possible that culturally determined bottlenecks are openly discussed;
   • To emphasize the importance of the experiments that are undertaken on the farms, and
   • To make people aware that they themselves can solve the problems, using their own technical knowledge.

In the KWDP, the review of the existing forestry practices within the district indicated that fuelwood trees had to fit existing land use practices. If fuelwood trees were competitive with pole production, women would find it difficult to access trees. The planting of trees by women was not common in the district. It was logical, therefore, to minimize planting by choosing species with coppicing characteristics. The stem of the tree should be crooked, so that it was not regarded as a pole product, and thus under male ownership. If possible, it should have many small branches which was preferable to one large trunk which later required splitting. These fuel-stick trees would have additional advantages as sticks burn more efficiently in open fires than large split pieces. In order to have a rapid impact in the area, the species needed to be fast-growing producing seed within one to two years, to generate a self-sustaining seed supply. These fuelwood species would also be used for other purposes on the farm but, initially, the project would concentrate only on fuelwood. The assumption was that once the trees were grown, farmers would find other uses for them.

The fuelwood species had to be produced in the improved farm nurseries but, to develop those nurseries, the local forestry authorities had to be convinced of the viability of the approach. It was not an easy task. The local forestry authorities denied the nursery knowledge of the farmers insisting that proper trees are produced in plastic polythene bags and not in bare root systems. But, from the perspective of the farmer, the bare root system is of more interest than the polythene bag. The farmer raises and plants trees on his or her farm in contrast to the forester who raises the trees in special nurseries. Box 13 looks at the advantages and disadvantages of traditional and farm nurseries.

Of particular importance, was the effort to create a better understanding of the fuelwood problem within the household. The fuelwood shortage was mostly felt by women who were not able to discuss this issue at family-level. The situation could best be described by acknowledging that a transition situation was underway in which the families were going from the collection of fuelwood to the production of fuelwood. It was essentially a kind of delayed neolithic revolution where woody biomass production systems were taking on the characteristics of the food and animal production systems. Men and women played a role in this process. There was a gender division of labour with women procuring wood, while men raised and planted trees on the farm.

Each village had its own tree nursery system. The techniques did not generally differ, but there was little exchange of plants between villages. Most tree species were distributed by the farmers who brought seed from other places to try on their

71

own farms when they came home by public transport or local taxi, called *matatu*. This *matatu* distribution system really existed. The challenge was to use it to spread the seed and fuelwood knowledge in the area.

To use the experimental capacity of the farmers, experiments were designed for farm level. The nursery system was chosen as the focal point of experimentation. Each farmer could start such a nursery without major investment.

A number of farmers were identified, some of whom had taken part in the earlier surveys. Discussions with these farmers formed the backbone of the participatory process in the project. Technical and extension ideas were discussed and modified by different local areas. For example, in the beginning two different farm nursery techniques were used, one for the north and one for the south of the district. In the north, farmers used the site of old houses as nursery beds because the old floors were very fertile. In the south, the farmers made nursery beds with cow manure. The work with the farmers clearly revealed that, from an extension point of view, only one system need be used. The system chosen was that of the nursery bed, to which manure had been added. A major advantage was that this system was already well-known to many farmers as it was used to produce their vegetables.

Another important step was made at the level of participatory technology development. Traditionally, to hand over new agricultural technologies, groups are formed to reach as many individual farmers as possible. In forestry work, an additional element is needed. Forestry nurseries need investments which are often too high for an individual. Groups are often, therefore, used to run a small forestry nursery and foresters then provide the farmers with important inputs. They train these groups on the assumption that, after a certain period, the nursery should run on its own. In practice, these nurseries collapse as the supply of inputs stop. To avoid this situation, the groups were not so much used to create a community nursery, but to learn how to set up a nursery on the farm of each group member, using local inputs available there.

If the project succeeds in mobilizing the farmers, seed supply can become a major bottleneck. From the surveys, it was clear that farmers were interested in experimenting — often trees with flowers are found around the houses while other species are planted as hedges which do not interfere with other crops on the farm. It was important to set up such a tree planting system so that new trees could be seen. For the project, activities had to be easy to establish and manage, but they had also to serve as demonstration areas for the new species and supply seed. The outcome was a seed production unit (SPU) a plot of land of 10 x 10 metres (later 20 x 20 metres) divided into four equal blocks planted with four different tree species at a distance of 0.5 x 0.5 metres. For the project, the SPU was an important object to show progress to the local forestry authorities.

This work was accompanied by a mass awareness campaign. Central to this was the commissioning of a play in the local language which explored the impact of fuel shortages on a household. The play attracted much local attention and audience participation and was the vehicle for launching extension comics — again in the local language — to drive home the message of farm-level nurseries.

The main focus of these awareness programmes was to create a realization that, based on their own technical knowledge, the fuelwood problem could be solved if they were prepared to change their attitude to tree-growing in general. It was based on two assumptions: first, that people did not talk about the fuelwood shortage within the family, and second that, brought into the open by discussion, the fuelwood shortage would be solved by the families themselves. This would, of course, require product development from trees so that local people could see real gains from planting.

Put simply, for wood products to be produced within the farmers' woody biomass production systems, new tree crops have to be developed which fit better into the existing land use systems. The experiments described in this project are the beginnings of such an approach.

Step IV: trying out

In this phase, the rural population start to try out the trees and discover new uses for them. The trees become part of the farming system and establish a place in cultural life. The awareness programme continues. Problems are presented in a way adapted to the local culture. Various ways to do this can be tried out. Discussion should be actively promoted. Exchange of experiences should be promoted.

In the KWDP pilot areas, two different approaches were followed. In one area, a play was developed on the basis of stories picked up during the discussions with the farmer groups. The storyline shows, within one family, that the woman is unable to collect sufficient wood one day. Her husband comes home and discovers that his chair has been demolished. The woman explains this by saying that she used the wood to heat the water for his bath. The man and woman have an angry discussion and the wife is forced to leave the compound. She goes to her friends nearby and explains the situation in her house. The other women admit that they also find it a problem to collect wood and start thinking about how to solve this. At this point a social forester enters the scene, he tells them special trees exist which are easy to plant and quickly produced. He also talks of small seedbeds, similar to the vegetable seedbeds the women make, that can be their own nurseries. The story ends when the women start talking about such new trees leaving the end of the play to the audience to join in the discussion. The characters in the play were people familiar and popular in the old stories from the district. At the beginning and end of the story, songs are performed about fuelwood problems and solutions. At the end of the play, seed and farm nursery pamphlets are distributed among the crowd. The number of farmers who came to the play and to whom seed was distributed was recorded.

A different approach was used in other districts. Large numbers of farmers were gathered at a rally to which important local leaders were invited. Songs and dances were performed around the theme of fuelwood. In one district, schools were heavily involved the school compound was the SPU site and farm nursery beds were made by the students. On school open days, parents were taught about fast growing tree species and introduced to the project.

As a result of these activities new species became familiar and in demand. This was particularly true after the first wood was harvested for fuelwood, when the product was clearly a benefit. From this point, processes started to emerge. In one of the zones, for example, an older woman was seen explaining the techniques of seedling raising and outplanting to her neighbours. The project had never asked her to do this but discovered it by accident when visiting her one day. All her neighbours were supplied with seed from the trees she had planted on her farm a year before. The seedlings were from her own small nursery which she established after seeing the play.

In another zone, the project officers learned that one woman, who had recently married a man from a village further away, had received as a present from her mother *calliandra* seed to secure her own fuelwood supply in her new house. In two years, the trees had established a good name as fuelwood source — it was a new wedding present!

It is obvious from driving and walking through those areas where the project was active that the trees have been adapted by the local population. One could see numerous trees, especially *calliandra*, on the farms, planted in hedges or around the house. The older ones had already been harvested and young sprouts were starting to grow. The wood, as fuelwood, was highly appreciated but several other end uses could be observed. Small poles were used as rafters in house construction and the leaves were given to the cows as fodder. Farmers adapted the species to their own local needs despite the project only talking about the use of trees for fuelwood. They responded to the challenges of the project and discovered, in their own way, a multi-purpose tree species.

Step V: sharing results with others

The underlying philosophy of the participatory development approach is that farmers are taking the initiative to improve their own conditions from inputs available to them. These initiatives are irregular and very difficult to recognize and to co-ordinate. The structuring process of local initiatives is the key issue for development, but there is little experience available to handle such situations or to train local leaders in such a process of leadership. Too strong steering makes the process rigid and stops development; too loose a structure creates the impression of chaos and poor management.

In order to communicate the basic ideas and principles of a participatory project, one has to agree on how such a process looks. Within the KWDP project, two different schools emerged, one demanding more time for experiment and the other arguing that it was time to share the results with other people involved. The first group focused on a traditional research model which meant that, only after strong indications were found, could the next step be taken. The second group favoured a dynamic approach. It was willing to court failure by pushing immediately into farms.

Step VI: sustaining the process

Working in isolation prevents awareness of how much development has occurred. Project staff concentrate on an inward-looking development process. It takes time to realize how much they contribute to others working in the district. The focus must move from the project staff being in the centre of change, to allow other people to become the centres. Instead of leading the process, staff have to realize that they can only guide the process.

At present, the project in Kenya is developing resource material for other officers in the district to be used in the training of trainers. All major institutions, including the rural development units, have been involved in the process which has created a consolidated network of knowledgeable people.

Wood beyond forestry: wood in rural development projects

With the methodology of participatory forestry development in mind, it is worth returning to why rural tree planting has faced so many difficulties. There are a number of bottlenecks in rural development programmes that make tree planting a serious issue. These are:

- A considerable lack of awareness about the potential of trees. Rural development workers do not know what can be done with them and what they can contribute to different sectors of rural development
- A lack of knowledge on how to start tree planting activities. Again, rural development workers do not know what species to use, where to get seed, how to produce seedlings, which techniques to use, where to get information about tree growing or what approach to use to involve farmers
- Planting of trees and production of seedlings are considered by non-foresters to be very difficult activities which can only be executed by highly specialized personnel, and
- It is often assumed that tree planting activities will lay heavy claims on financial resources, staff and logistics. The images of large tree nurseries, the transport required for seedling distribution, the large areas to be cleared and planted, staff to be trained, and foresters and labourers to be employed sometimes creates a fear that tree planting would disrupt planned activities, especially in agriculture, at a time when resources are scarce.

What is needed is a method which enables project managers to introduce tree planting activities in phases without laying excessive claims on staff and resources. Some outside assistance may be necessary to guide and stimulate the process. However, this does not necessarily have to be a large input in terms of time and money.

Again, a cautionary approach is advocated based on the following principles:

- Any tree activity must be based on insight into the present and potential role of trees and shrubs in rural areas, the importance of trees and shrubs for the rural population, the knowledge and experience of rural people, and the potential of existing and new tree planting methods and techniques
- The approach should be phased, starting with small, easily manageable activities that gradually become an integrated component of the project. During every step, the programme can be adjusted, stopped or accelerated, depending on the results of previous steps and the possibility for expansion, and
- In the approach, distinct phases containing trials, training, monitoring and evaluation are necessary. This enables continuous improvement and reformulation of the activities and also enables different projects, carrying out similar activities, to compare their achievements from time to time in a systematic manner.

The activities should, as much as possible, be self-generating where relatively small input can have a large impact. Methods to achieve this include the training of higher and intermediate staff who in turn train field staff, working with individuals or groups of farmers who pass on information to others using techniques that are accessible to the target group of farmers.

New methods and techniques are more likely to be adopted by rural people if they are close to what people are doing already. By using low-input methods, large sections of the population can be reached, not only the easy adopters but also the resource poor groups. Local practices are a starting point for a dialogue with rural people about the improvement of tree management. Using local initiatives, the focus of attention can be shifted from the provision of inputs to extension activities focusing on tree management.

Rural tree planting is still a new area of work and knowledge about agroforestry production systems is often limited to specific conditions or regions. Off-the-shelf packages are not available. This means that development work has to be done on the ground. The new activities should be set up so that no dependency on outside expertise is created and so large additional investments are not required.

Many rural tree planting issues are influenced by, and have influence on, other components of land use. It is important to realize that agroforestry development cannot take place, in the long term, in isolation from other development sectors, in particular from agriculture and animal husbandry. It is also important that co-operation is sought with all relevant institutions that work in the same rural area. It is necessary quite simply because there is little money available for rural -development. Integration is needed to avoid duplication or counterproductive competition, to exchange information and support each others programmes, and to draw a consistent line of action towards the farmers.

If linkages can be established, there is no need to establish separate rural tree planting extension organization. The agricultural extension service can spread tree planting messages on a large scale even when they have been developed by foresters in small pilot projects.

There is, clearly, no need to begin separate rural forestry projects if tree planting can be integrated into other projects. However, some general guidelines are useful for the development and integration of tree planting activities into existing projects. The integration of a forestry component into existing projects is a step-wise process which enables local people to control the process, follow the progress and adapt the programme in a gradual way without deviating from the main project objectives.

There are five phases which are distinguished in the integration process. They are indicative; in reality there may be more or less. The phases are largely based on experiences in Kenya, where tree planting was integrated into rural development programmes, especially the National Dairy Development Programme.

Phase I: situational analysis

Any initiative to develop something new has to make use of the potentials that are present within the present agroforestry activities carried out by the rural people. It must also deal with the constraints of the present system. These constraints can be of a different nature. The constraints can be related to the knowledge base of people or project staff but can also be of a physical, institutional or political nature. Recognizing these constraints, and analysing what can be done to remove or reduce them, is the important first step.

In order to begin a situational analysis project staff must look into the following issues:

- The present function of woody vegetation for rural people in relation to their priorities, and in relation to the project objectives
- The present role of rural people and the relevant institutions in the management of woody biomass
- The present problems or shortcomings in the woody biomass system, and
- The present efforts being made by rural development institutions in the field of woody biomass development.

The list of questions in Box 23 may be useful to start this analysis. Such an analysis does not have to be a major study that takes years and a team of experts. It is only important to know the key issues influencing woody biomass development in qualitative terms. Since the information obtained in this phase will be used in the following phases, it is important that the analysis is carried out by staff who will actually be involved in the following phases.

No detailed surveys are necessary at this stage, but discussions with the main actors playing a role in woody biomass management are a must — farmers, foresters, agriculturists, rural sociologists and policy-makers all have an input. It is important to gain insight into technical, socio-economic and cultural factors influencing woody biomass management. It is therefore advisable to carry out the analysis in a team with a wide range of expertise.

Phase 2: identification and formulation of agroforestry options
During the initial phase of identification, a number of promising agroforestry options have to be identified and formulated in such a way that the programme can start testing. Both the existing knowledge of the farmers and external information have to be employed. During each of the following phases, the

---

Box 23 Five questions to organizations involved in rural development

1. **What is your present contribution to woody biomass development in the area?**
- Forest management
- Rural tree planting activities
- Tree seedling production
- Extension work on tree planting
- Mass awareness related to tree planting and conservation, and
- Research on forestry or agroforestry, on-station or on-farm.

2. **What is the policy of your organization towards woody biomass development in rural areas?**
- Specific objectives
- Target groups
- Target areas, and
- Approach.

3. **What are the major constraints and limitations to reach your objectives?**
- Resources
- Management
- Physical
- Political, and
- Knowledge.

4. **What do you see as the major role of your organization in woody biomass development in the future?**

5. **What do you think could be the role of the rural people?**

agroforestry methods are reformulated and refined on the basis of monitoring and evaluation results.

Since testing of tree planting options starts on a limited scale, it is not wise to address all problems and priorities identified in phase 1. It is better to identify some specific focal points. These focal points are obvious including:

- Relevance to the project goals
- Relevance to the target groups that have been identified by the programme, and
- Important enough to the people concerned.

Forestry programmes often have a wide scope so a number of different focal points can be chosen. Examples of focal points can be fodder production, tree seed supply, production of building material, and the demand for specific tree species or firewood production. At this stage, it is best to select only those issues that do not require drastic organizational, political, legal or financial changes.

The formulated options must be concrete and clear, easy to implement, cheap, and easily adoptable. All the people involved should, as far as possible, agree with the decisions that have been taken, and then start with a bang. The choice of options will have implications for the organization. It may involve changes in the organization including tasks of staff, supervision of nurseries, and collaboration with other organizations. It may also involve changes in extension by refocusing target groups, communications with farmers, publicity, technical messages, seedling and seed distribution.

The involvement and training of project staff is important in this phase. It lays the foundation for the involvement and motivation of the staff in the implementation of activities. Information and training sessions provide a forum for the exchange of ideas. At this stage, emphasis must be given to the following subjects:

- Understanding the role of trees in a rural context
- Recognizing and identifying farmers' knowledge
- Understanding the process of formulating agroforestry options
- Establishing and maintaining of trials, and
- Monitoring and evaluating procedures.

Phase 3: Testing

The selected options are tested to find out which of the chosen agroforestry methods qualify to be further incorporated in the programme. The result of this first round of tests can not be a clearcut 'polished' system, ready for wide application but more a rough set of ideas that can be developed further.

Some characteristics of the test activities are the following:

- Testing is done on a small scale, in few locations, on a limited number of farms
- If possible, trials should be established on selected farms and in experimental plots
- The set-up and implementation of on-farm trials is done in close collaboration with farmers, and
- Quick monitoring and evaluation methods are used.

An important decision to be taken during this phase is how are the trials going to be established. Are seedlings going to be used or seed? In the case of seedlings, who is going to raise them? Is seedling or seed distribution the way forward? If seedling distribution is chosen, what type of nursery is required?

Training activities during this phase focus on the results of the trials, the reactions of the farmers, and the consequences for the next phase. Monitoring and evaluation procedures are also discussed.

Phase 4: partial integration

During this phase, agroforestry systems are reformulated on the basis of the results of phase 3 and are further developed within the structure of the programme. The result must be a set of agroforestry systems that can be integrated fully into the programme.

Agroforestry methods are tested on a larger scale with more farmers, while additional on-station research continues. There is a larger variation in species and management systems and the activities become gradually part and parcel of the on-going programme. Monitoring and evaluation provides information for further research on species and management systems.

The activities are becoming a more or less integrated component of the programme. The main question monitoring and evaluation activities should address is whether the agroforestry options are good enough to be taken up in extension programmes to be run by the project or other extension agencies.

Phase 5: complete integration

By now, agroforestry should have become a fully integrated component of the programme. The results of the work carried out should be made available to a wider audience for application elsewhere. This implies that the following activities are carried out:

- Formulation of agroforestry extension messages for large-scale dissemination
- Incorporation of the packages into extension programmes
- A broad range of species and management systems are offered or under study
- Continuous monitoring and evaluation
- Continued training of project staff
- Further research on species and management systems, and
- Handing over of the results to local organizations by way of training, and demonstration plots.

The National Dairy Development Programme - an example

So far, so good. There is a methodology, but does it work? The case of the dairy programme in Kenya, provides a useful example. The National Dairy Development Programme (NDDP) aims at increasing milk production in medium- and high-potential areas. It operates in 14 districts and attempts to introduce improved zero grazing methods to small-holders. Over the years, NDDP has concentrated on the use of napier grass for the bulk of roughage production. In spite of the good results that were obtained with this crop, it has one major disadvantage: its relatively low crude protein content. Good milk production can only be obtained if napier grass is supplemented with protein-rich feeds, such as concentrates, other roughages or by-products.

Fodder trees provide interesting opportunities for the NDDP. Many tree species have a much higher crude protein content than grasses. The trees can provide valuable additional fodder during dry periods when other feeds are scarce. Additional fodder from trees can improve the overall quality of fodder, not only the nutritional value, but also digestibility. Furthermore, if the roughage bulk is fairly secure, only small quantities of tree leaves are required to achieve a considerable increase in milk production. As most farmers in the Kenyan Highlands are familiar with tree planting and many farmers have experience with indigenous fodder trees, it was obvious that forestry could be integrated into the NDDP Project.

During the first year a number of activities were carried out. These centred on the training of NDDP staff. The training focused on an introduction to the potential of fodder trees, an analysis of fodder situation in the districts and technical information about possible fodder species. Field days were used to observe aspects of traditional and modern fodder tree usage and tree planting activities by farmers. District plans were formulated for the initial testing.

In each district, a number of farms were selected for trials with a couple of fodder species. On 275 farms, hedgerows were established, each having a length of 15m, either established through direct sowing or by planting seedlings from an on-farm nursery. Within a year the first fodder harvest was expected. During the first year the trials closely monitored the survival rates and growth rates of the trees, fodder yields of the hedges, and the increase of milk production as a result of additional tree leaves.

The species selected for fodder trials were *Calliandra calothyrsus*, *Leucaena leucocepholia*, *Sesbania sesban*, and *Gliricidia sepium*. The major advantage of using these tree species was the experience with the trees in other projects within the country. Fodder from trees can be harvested in a most optimal way if trees are planted in hedges and regularly cut. The production of wood is suppressed because of the cutting of the leaves of the plant which are harvested as fodder. Production figures from elsewhere indicated a reasonable production potential of about 1-2kg dry matter for each metre of hedge densely planted. The system of fodder hedges was therefore chosen to be introduced on the farms.

After 18 months, a follow-up was undertaken of the activities initiated earlier on. The results after this short period were encouraging, particularly because farmers had extended the length of their hedges from an initial 15 to 60m.

It is possible to build agroforestry in other projects using the methodology outlined in this chapter. Certain things must be remembered. First, the different components are strongly interrelated. Second, the methodology is a step-wise, logical process. Third, the methodology aims at minimizing the need for outside assistance. Agroforesters only play the role of catalytic agents, to set the process in motion. Fourth, the methodology must mobilize local agroforestry knowledge. To this end, intensive contact should be maintained with farmers. Finally, the training of project staff plays a crucial role in the methodology not least to encourage project staff to abandon professional bias.

During each phase, a training programme is established which covers the experiences of the previous period and looks ahead to the coming period. In the beginning, the emphasis will be on farmers' knowledge of trees and tree management. Later the emphasis will shift to trial methods and procedures, the formulation of extension messages and monitoring and evaluation methods.

Monitoring and evaluation are important elements of every phase because any effort to improve or redefine what has been done already should be based on reliable information about performance. This information should analyse the overall context in which the introduction of agroforestry methods takes place, the inputs and outputs of the system and the acceptability of the methods for the farmers. Above all, the farmers' opinions should indicate the direction in which the programme should move.

The implementation of the formulated interventions is the responsibility of the project and should be carried out in such a way that the new activities do not lay heavy claims on the on-going activities. On the other hand, it must be realized that developing anything new requires time, resources, people and a lot of patience. Therefore plans for new activities should be realistic, flexible and iterative.

It is possible to include forestry in non-forestry projects — a so-called cuckoo's egg solution to building wooded landscape. But to do so requires rethinking existing projects and adhering to some broad guidelines. These are essentially guidelines that say; farmers first; share information; and start small. Above all, the guidelines say experiment.

6. A New Approach to Training

In the first part of this chapter, modalities for technical assistance, training, planning and field staff involvement in project programming are presented. In the second part of the chapter some practical issues regarding the organization of projects are analysed and some suggestions for the way to solve these problems are presented.

New modalities: starting points

So far this volume has argued two critical points. First, there is a need, and a challenge, to adjust the classical forester's attitude so that he or she can begin to tackle the problems of rural forestry. Second, this adjustment must recognize the range and value of local people's experience where they protect, enhance or build new woody biomass landscapes as an integrated part of their livelihood systems. By itself, this recognition is not enough. A participatory development approach that will build a sustainable biomass programme by involving the forester with local people is needed.

Already, there have been hints about how to tackle the problem. First, there is need to turn the administration process inside out, to reverse the normal delivery channels for forestry inputs, to make the priorities of intervention, and thus research, be community, not professionally, led.

Most present planning systems are complex. Many planners talk about the involvement of the local population in project formulation. Plans, however, are made to a different timescale, to a different life cycle than the seasonality that dominates farmers. Planning also means different things to different people. For example, a donor organization wants to read from the plan the number of trees to be planted each year in order to calculate the impact of the project. But the local population, that participates in the project, wants to read the plan to see what it receives, individually and communally, from its commitment to the project. These different perceptions have to be recognized by staff formulating the project.

Projects are often formulated on the assumption that farmers' knowledge is limited and, consequently, outside knowledge has to be brought in to the project. Local knowledge is not acknowledged and does not form an input into the project, as a result, most forestry projects are identi-kit, checklist interventions. A typical project plan entails that tree nurseries are set up in the project area, seedlings are distributed as soon as the rains start, then labour from the farmers is enlisted. Thus, in professionally led projects, the level of participation depends on the organizational talent of project staff, not on the commitment or motivation of the community.

To reduce the danger of this happening, it is important to start with a process of knowledge exchange. This can be done by organizing occasions where professionals have opportunities to exchange ideas and to learn from each other, such as training courses, policy workshops and seminars. But, most importantly, the individual experience of participants needs to be given an explicit role in these training courses, seminars and workshops. An approach which highlights the role of the individual will stimulate the participatory process in training.

The major emphasis of the information and knowledge exchange should not be on finding an overall solution for certain problems, but rather on the high range of solutions available within the various local circumstances. Projects should neither try, nor have the desire, to produce universal technical guidelines. The differences in approaches with regard to forest and tree management and agroforestry techniques in each country, department or region are so extreme that general technical guidelines will always be superficial.

Projects should be designed to stimulate the development of systems and channels to let the information flow to those who need it most, that is field staff and farmers. Projects should activate and stimulate the knowledge exchange of those directly working in the field by creating information flows reflecting the problems as perceived by field staff and farmers.

Modalities for planning

It is impossible to formulate a universal planning methodology. Such approaches not only lead to damaging blanket approaches, but also to subjective planning processes. Rather, approaches should be attempted which have a high level of objectivity. Technical assistance projects should stimulate such a process and investigate which of the existing procedures would be best adaptable to the given circumstances.

What is needed is a management tool which facilitates planning, execution and evaluation of a project. Such a tool serves as:

- A format for presentation to donor and partner authorities: project ideas, preappraisal reports, project documents and progress reports
- A summary of the project in the form of a matrix that remains valid during project implementation but can be modified. The matrix contains information about development objectives, immediate objectives, and outputs with indicators to match, inputs, and external factors necessary for attaining the development and immediate objectives or the production of the outputs, and
- A sequence of analytical tools which is used in an external/internal workshop situation.

One such method is the logical framework approach where affected groups, problems, objectives, strategy, project elements and external factors.

Modalities for training

A similar process should be followed in training. Most present training methodologies are based on a one-way information flow. It is well established: a teacher tells the class what to do. Most technical solutions, however, are never screened for, for example, adaptability. To develop objectively oriented training standards in isolation should not be the aim of assistance projects. An important output of objectively oriented planning and training approaches must be a much more defined role for the participation of the people living in the region to be covered by the project.

The problem with forestry training is that it is based on teaching people about trees. The new model of forestry training must be based on people, not trees. An emphasis on trees is essentially an emphasis on techniques. Such an emphasis produces a focus on things. The new emphasis on people is an emphasis on development. As such it emphasizes not things but relationships — relationships between people, between people and trees and the landscape they live in.

In general, a systems approach to forestry training can deal with techniques alone, but it is rigid and cannot encompass the range and variety of human experience found in the landscapes in the field. It operates at an abstract level which, by definition, is non-operational.

If, however, we are interested in the development of people and their landscape we have to understand the relationships between people and between people and their environment. To understand the relationships we have to study the on-going process for which we need a framework. The framework is essentially a framework of questions. Even establishing this framework, we must accept that answers to the questions will be different depending upon people and place. The framework consists of guiding questions rather than answers.

To teach in this relational manner requires that we start from a series of questions about the professionals themselves rather than the subject. Three key questions signal the point of departure. We have to ask:

- How do you think?
- How do you work?, and
- What do you observe?

These basic questions, if applied, have to be focused on the subjects one wants to teach. For example, if community forestry is the subject the question How do you think? can be translated into What are, according to you, the three most important objectives to community forestry? How do you work? can be translated into a task to be performed by the students. A plan can be made on the basis of inputs provided in a handout. And What do you observe? is translated into the task of making a drawing from the farm from which they originally come. All the answers are finally compared and synthesized by the professionals. Often this exercise shows knowledge gaps to the participants and these gaps create the interest for further learning. It produces the answer to the following critical question:

- What do you want to learn?

In such a learning model, the process of acquiring a skill is centrally defined by existing personalities.

One of the most important issues for a person in a learning process is acquiring sufficient self-confidence for action. This self-confidence for action is based on him or her discovering, in his or her own terms, what is right and what is wrong.

The new approach, based on people, is essentially different from traditional training methodologies. Some of the differences are summarized in the table below that contrasts the properties of traditional education and education emphasizing people.

Traditional	Emphasis on people
Conforming	Changing

Control	Liberation
Authoritarian	Democratic
Passive	Active
Vertical	Horizontal
Academic	Popular

In the new approach, the notion that facts are taught must be rejected. Instead, experience is encouraged. Everybody has experience so everybody knows something. By emphasizing this, the participant gains self-confidence. Such experiences form the basis for a possible co-operation between the peasant and the social forester at a future date.

The new approach can be applied at several levels:

• Teaching new foresters
• In-service forestry training at the institutional level
• Foresters working in the field with farmers
• Foresters working with fellow colleagues in agriculture and husbandry, and
• Foresters working with teachers at primary and secondary schools.

Modalities for field staff involvement in project programming

It is one issue to identify what should be done differently, but it is an entirely different matter to determine who can do it. Clearly the staff of a development assistance project can play a catalytic role, but how do they work at the field level and with whom to understand and tap the local knowledge base? How is access to that knowledge base gained, and once gained, how can the new insights be enhanced and then returned and institutionalized on a broad basis?

A significant gap has historically existed between the opportunities at the field level and the intentions at the project level. The top-down approach has been proven to have limited utility by itself. The bottom-up approach has suffered from communication gaps and a lack of genuine understanding.

To help bridge this gap, a new and under-utilized resource has emerged in the past 10 to 15 years, namely, national field level staff who are highly professional, experienced, motivated and positioned to understand conditions and needs at the grassroots level. The participation of the experienced field staff in identifying problems, proposing solutions, recommending priorities and effectively accessing the grassroots level in terms of needs, opportunities, and knowledge acquisition and dissemination, is indispensable. All of these field staff have significant professional obligations in their national or departmental settings. Consequently, requests for their help in developing projects must be highly targeted and efficient. Small working groups can serve as technical advisory committees through occasional workshops and other mechanisms. The committees must be composed of staff who are peers in terms of experience and level of understanding to facilitate maximum efficiency in terms of group dynamics.

It is suggested that meetings with these professionals have to be:

• Structured in terms of purpose
• Open and informal to encourage a lively and uninhibited exchange of views and ideas, and
• Serious commentary on project plans.

By means of a properly guided process, the group dynamics of work sessions with professional field staff in a free environment can yield new insight and knowledge that would otherwise go undiscovered. These advisory committees are not suggested as a substitute for national co-ordinating committees who necessarily serve as policy advisers to technical assistance projects.

To summarise the new modalities a number of points can be drawn out:

1. The emphasis should be on project modalities that will have maximum effectiveness in the development of new insights as a result of improved access to, and understanding of, the enormous knowledge base at the grassroot level.
2. Outputs should include objectively based planning and training methodologies and approaches that capitalize upon the rich variety of experiences and needs of the participating countries.
3. The goal of a project should not be to homogenize approaches to planning and training, but rather to apply objective methodologies that accommodate and build upon the rich diversity of needs and characteristics of the region.
4. Systems should be developed that permit the diversity of knowledge and experiences in a region to be widely and readily shared.
5. The utilization of professional and experienced field staff to assist with problem analysis, identification of solutions, recommendations on priorities and in gaining access to the local level is a necessary, but not sufficient, condition for success.
6. Co-operation and co-ordination with other donors and organizations that provide technical assistance is needed, and
7. In a project context, there is need to:
   • Strengthen human resources
   • Support networks for the exchange and sharing of information
   • Provide advice on technology assessment and new technologies
   • Provide advice and design services for institution building, project formulation and curricula, and
   • Support workshops, seminars, the production of newsletters and other activities as identified through the project problem analysis process.

Practical issues

This book promotes a participative approach towards farmers. It is, therefore, necessary that the organization of a project is structured in such a way that there is space for ideas to come from the bottom up. This is not an easy task, because it entails a complete reversal in the contemporary development paradigm. The transfer of knowledge has traditionally been from the top, that is the professional level, to the bottom, that is the grassroots level, with a blatant disregard for anything the farmer has to say. Rethinking and redesigning this process will take time and will not always be smooth running. Successful projects are essentially defined, not by what is going right but how people are dealing with what is going wrong, that is how effective the participative, bottom-up approach is in solving problems.

For strong organizational structure, it is not a matter of defining rules to make things right, but to suggest a framework of rules to handle what is going wrong. An organization should not be built on the supposition that everything goes well, but it should focus on the design of mechanisms to be used in case anything goes wrong. Usually rules are used to define what is unacceptable, what are the boundary conditions. Successful people, successful projects and successful institutions are usually built the other way. They all operate with the sense of relation to define what can be wrong at the core.

Another point to consider under the heading of practicalities is the introduction of knowledge from outside of the project. It is very important to use all available natural resources of the community. For example, when the community possesses many shrubs, production of honey by bee-keeping may be an option. However, a forester, professional or extension worker, generally has little knowledge of bee-keeping. The structure of the project should be such that expertise from outside (usually local) can be invited to co-operate in order to compensate for this omission. The most important factor is to mobilize local knowledge. Apart from intensive contact between project and farmers, the project should also enhance the exchange of experiences between the farmers.

The underlying philosophy of the participatory development approach is that farmers are taking the initiative to improve their own conditions from inputs available to them. These initiatives are irregular and very difficult to recognize and to co-ordinate. This structuring process of local initiatives is the key issue for development but there is little experience available to handle such situations or to train local leaders in such a process of leadership. Too strong steering makes the process rigid and stops development. Too loose a structure creates the impression of chaos and poor management.

Project staff are needed who can stimulate the on-going process without directly needing to lead it. People are needed who can keep apart from the actual activities without losing interest in the detail. Each individual needs a level of total independence to make local decisions which are important to steer the process. Unfortunately, given current management practices, most personnel are selected on the basis of their expected way of operating within a team. This is often determined by the capacity of a team leader to handle different characters, which is hardly the management model for participatory practice.

In many projects, the personnel who actually have most contact with the farmers are the lowest in rank and therefore selection is not considered to be important. The personal qualities of an extension worker are probably decisive for the success of the co-operation between a project and farmers.

There are circumstances in which it is impossible to select personnel. One simply has to work with the personnel already available. Their qualities can be improved by giving adequate training, but some people are simply not fit for the job of participatory development — some people are not 'good' with other people.

When new personnel have to be contracted, a selection can be made. It might be useful to develop criteria for the selection of extension workers and other project staff. Some of the most important criteria are:

- Empathy for the way people learn and are motivated
- Willingness to use own initiative
- Able to speak the language of the farmers
- Prepared to work under circumstances with very few facilities, and
- Possessing good communicative skills.

It has to be observed, again, that technical qualities are of secondary importance. If an extension worker does not know a certain technical issue, he or she can ask other people within the project. However, if the extension worker does not have good communication skills, it will be a serious limitation.

One example demonstrates this. Cajamarca is a department in the north of Peru where the Peruvian Community Forestry project began very late. At that moment, no extension material was available for the new region to support the extension workers as they started work in the villages. In one village, the intervention of the extension worker was very peculiar.

At his first visit he told the local peasant organization, the *ronda campesina*, what he wanted to do. The president of the *ronda* was not very enthusiastic but he agreed that the man start work. Every day the extension worker visited a peasant of the village, walked with him along his fields and discussed the erosion problems, the ways soil conservation techniques and trees might solve these problems and the way trees might improve the quality of the soil or provide other benefits. After five months, he had talked extensively with everybody, but not one seed or one seedling was sown in a community nursery.

He organized a meeting with all the members of the *ronda* and proposed to start a community nursery. He had gained complete confidence of the peasants and a community nursery was started. Five months later, 11 months after his first visit to the village, the nursery contained 11,000 seedlings, more than in most of the villages in other departments in which the project had intervened for more than three years. At that same time, the man had organized a women's club and, in close co-operation with the women, had installed some improved woodstoves in their houses. The villagers were that eager to co-operate with the project and to increase their knowledge that they wanted to be the host to a course on agroforestry organized for all the extension workers of the Cajamarca department. This success, of course, cannot be contributed solely to the extension worker, but it is clear that his intervention was very important. The extension worker had only recently joined the State Forest Service, but possessed many years of experience in villages, working as a school teacher. He was used to listening, not instructing.

Extension material has an important role to play in projects: it is used to disseminate certain themes, techniques or methodologies. Extension material might be used in the following circumstances:

- To promote a discussion with and between the farmers
- To emphasize the importance of experiments that are undertaken by the farmers in co-operation with the project, and
- To make people aware that they themselves can solve the woody biomass problems, using their own technical knowledge.

In many projects much extension material is produced to raise awareness in general. Raising awareness is only useful in specific circumstances. In situations of real scarcity, raising of awareness is hardly necessary. When ample forestry resources are available, there is no interest in forest protection or tree planting, and raising of awareness is of no use.

During the presentation of extension material in villages, many workers only show a video or filmstrip and do not promote a discussion afterwards. Extension material is not a substitute for discussion. It must not induce a consumption mentality among farmers. Active involvement of both the extension worker and the farmers is required.

If extension material is going to be used, it is important to think about the type of material to be used. There are many types of extension material, for example, filmstrips with cassettes, videos, sequences of sheets, posters, calendars and publications. One should choose the right medium for the specific goals to be pursued. For this choice, important factors to be evaluated include time of the day when the target group (for example, men, women, children, or the whole village) can best be reached (day or night conditions), clearness of the contents (for example, spoken language on the cassette or video, written material which will not be useful to people who cannot read), size of the group, attractiveness of the medium, reliability, transportability and cost. The role of formal extension materials for the social forester is almost certainly smaller than in many conventional extension projects in which extension material is actually used to push a certain message. The social forester tries to use local means of communication as much as possible such as gathering of people around village meeting points.

A programme which assists primary school teachers to present lessons on forestry, trees and environment may be of great help to the forester to promote the basic importance of trees. This seems a long run investment. However, most of the pupils in rural schools will be working on the land within ten years which is not a long time for an investment in development. Moreover, children talk with their parents about the things taught to them at school. Programmes directed to both children and adults may strengthen each other.

A primary school programme should preferably be designed according to pedagogical criteria. For each class, a programme can be designed in which important aspects like trees around the village, trees and agriculture, the various uses of the trees, lifecycle of the trees and tree protection are treated. It is also possible to direct a magazine at the school teachers. In such a magazine, common problems can be treated or new themes can be introduced with some indications how to treat the subject in the class. Special materials can be provided, such as school books for the pupils, games, and compilations of fairy tales in which trees play a role.

Even if nothing of the kind exists in a social forestry project the co-operation between the forester and the school teacher still can be very fruitful. School teachers may have a great knowledge about what is going on in the village and may be able to provide valuable information to improve the co-operation between the forester and the village. Many school teachers in rural areas suffer from a lack of materials. If the forester provides the teacher with information available to the project, many teachers will be glad to use it for their lessons. Some assistance in practical matters, for example an excursion with a class through the village fields, may be significant.

This chapter has emphasized the need to look for new modalities that support the implementation of a new social forestry. It offers few guidelines, only a sense of a way forward. That way forward demands as much experimentation in the classroom as on the farm. Such experimentation is necessary to build people's confidence so that they can build their own landscapes.

7. Conclusion

And now?

It is difficult to write a conclusion to this book. It has laid down a challenge to foresters, and other development professionals, to move from a new way of *seeing* to a new way of *doing* wood development projects. Along the way, we have emphasized that we must understand how people use wood, how they manage wood production in their local environment and their knowledge of wood regeneration. We have laid great emphasis on the argument that social, not technical, issues are the key to success. We have stated categorically that participation is a process that requires that control is taken by the local people not retained by professionals.

When we return to western Kenya, we can see, in the landscape, numerous individual trees and bushes that stand as monuments to the forestry activities of local people. On farms and along the road, these trees begin to tell landscape stories about how people have built the Kakamega landscape to meet current and future needs. The trees stand as monuments to local effort.

That is what has to happen now. Foresters need to go forth and explore what people themselves can do. Other landscapes will produce other numerous monuments as people build their livelihood and landscape systems. Success, with such a diffuse programme, will be difficult to measure. It is not simply a matter of spotting individual trees or even farm nurseries. Success is when local people do not complain about wood shortages. The challenge is to build landscapes to produce nature in such a way that local entitlement is met. Go out.